T0321122

Bio-inspired Networking

Series Editor
Pierre-Noël Favennec

Bio-inspired Networking

Daniel Câmara

First published 2015 in Great Britain and the United States by ISTE Press Ltd and Elsevier Ltd

ISTE Press Ltd
27-37 St George's Road
London SW19 4EU
UK

www.iste.co.uk

Elsevier Ltd
The Boulevard, Langford Lane
Kidlington, Oxford, OX5 1GB
UK

www.elsevier.com

Notices

For information on all our publications visit our website at http://store.elsevier.com/

British Library Cataloguing-in-Publication Data
A CIP record for this book is available from the British Library
Library of Congress Cataloging in Publication Data
A catalog record for this book is available from the Library of Congress
ISBN 978-1-78548-021-8

Printed and bound in the UK and US

Contents

Introduction

*Word cloud representing the full text of this chapter and
the words frequencies. Created with Wordle.net*

Even before the computational meaning it has today, the word
"network" was intrinsically linked to biological and natural structures.
The earliest occurrence of the word *network* in print media in English
language dates back to the Geneva Bible of 1560 *"And thou shalt
make unto it a grate like networke of brass"* (Exodus xxvii 4). Here it
refers to a grid of metal wires; however, according to the *Oxford
English Dictionary*, in 1658, it was already used to designate the
reticulate structures found in animals and plants. Later, in 1839, it is
introduced as a way to describe the relations among rivers and canals.
The very formation of the word is a juxtaposition of "net" and "work".
"Net" is an old English word used to designate a spider's web,

moreover the World Wide Web, or just the web, that some use to refer to the Internet refers also to the same spider's structure.

Nature has been a source of inspiration to humans for many centuries. We observe what nature has done and use it as a source of inspiration to solve problems in other contexts. This process is called *biomimetics*, derived from the ancient Greek β*ίος* (bios), means life, and *μίμησις* (mīmēsis), means imitation, or *μιμεῖσθαι* (mīmeisthai), means to imitate; thus, biomimetics is the imitation of life processes. The literature is full of examples where nature directly inspired innovation.

A particularly interesting case is Velcro®. Velcro's history begins with its creator George de Mestral, a Swiss engineer, who conceived Velcro in 1941. It was inspired by the way bur seeds attached to his dog's fur, and his own clothes, after a trip over the Alps. Examining the seeds, Mestral observed that it had small hooks and these could attach to a series of different materials. In fact, anything with a loop where the hook would fit. From that, Mestral perceived he could use this to easily attach and detach materials. Today, Velcro is considered as a key example of nature inspiring humans [VOG 88] and the way we can apply nature's mechanisms in other domains.

Another interesting example is how morpho butterfly wings have inspired the development of display technologies [WAL 07]. The interferometric modulator display [QUA 08], the basis of Qualcomm's Mirasol technology, is inspired by the microstructures that give Morpho butterfly's wings their color. Instead of simply reflecting the light, as any regular pigmentation, morpho butterfly wings use structural coloration, i.e. they have microscopic structured surfaces that interfere with the way light is reflected. These structures have successive layers and they repeatedly reflect the light in different and specific wavelengths. This results in vibrant colors due to a thin multilayer interference film and its scattering properties.

History is full of examples where nature has inspired people and their work. A good example is the Ornithopter, one of the most famous inventions of Leonardo da Vinci. By imitating and adapting the very same methods that nature uses in other contexts, the ornithopter reflects well this desire man has to go beyond the limits nature imposes on him. Even if Da Vinci's ornithopter never worked the way he had intended, it is clearly inspired by the flying characteristics of winged animals, especially bats. In fact, it took more than 500 years after Da Vinci's first designs for a fully man-powered flying mechanism to be built. In 2010, researchers at the University of Toronto at the Institute for Aerospace Studies were able to build, and successfully fly, a man-powered ornithopter, the Snowbird, which flew for 19.3 s.

Computers have brought us the ability to process large amounts of data and automate a series of processes. They have even made possible efficient communication over large distances through computer networks. However, we are always searching for methods to improve these characteristics and significantly decrease the human intervention in these processes, while improving the speed and agility of computer systems. The efforts in this sense can be either top-down or bottom-up.

The reasoning of top-down approaches is to get the broad view of the system and then look into the details, i.e. start from the user requirements and from that, derive the code implementation to solve a given problem. Methods that follow this approach are, among others, protocol synthesis, starting from a high-level specification [SAL 96], and the derivation of policy rules from high-level representations. The bottom-up approaches look at how high-level functionalities would emerge from the interaction of lower level units. Swarm intelligence, artificial life and evolutionary computing are examples of techniques that favor bottom-up kind of thinking.

While the top-down approach seeks a more formal way to describe and construct software, closer to the human mental model, in general, nature has a rather more bottom-up approach. Even the simplest life forms possess a level of robustness and adaptation far bigger than the current artificial systems. Considering these, even if sometimes it looks counterintuitive to us, it seems reasonable to learn from biology in order to draw inspiration for the design of new computer systems.

Nature's methods are the result of centuries of a continuous massively distributed trial-and-error process. The whole process is so vast in terms of time and number of attempts that it is even difficult for us to imagine and completely understand it. Even though we ignore the influences of man in the evolutive process, globally, hundreds of new species appear and disappear each year [GOR 00]. The survival of a given species is linked to its capacity to adapt to the environment and find a niche where it can evolve and reproduce. It is estimated that more than 99% of all species that ever lived on our planet are now extinct, most of them even before the arrival of humans [NEW 97]. Even more, half of the species that currently exist may become extinct by 2100 [TIV 12]. Understanding this process is important for many reasons, including our own survival as living beings. The world has already seen many changes, and a number of other changes will still happen. Equilibrium is an important concept in nature, every time a new and more suited species appears it influences the environment where it is inserted. This environmental change may affect other species, which need to adapt to the new conditions. This adaptation process will eventually reach an equilibrium point.

In general, stability is a desirable characteristic for both biological and synthetic systems. "Homeostasis" is the name of the property of some systems to self-regulate and remain in a relatively stable condition. The term "homeostasis" was first used to describe a series of processes internal to living organisms, e.g. body temperature self-regulation process. However, today, it has a broader usage; any natural or artificial system capable of self-regulation and having the tendency of converging to an equilibrium state is said to have a homeostatic behavior. In nature, we have a number of processes that present this predisposition. For example, the delicate balance between species in a given ecosystem is proof of this. An ecosystem, a main concept in biology and ecology, is defined as a set of integrated living beings interacting with each other and with the surrounding environment.

The predator–prey relationship is fundamental in most ecosystems. Predators have a major role in the equilibrium of the ecosystem; they help to regulate the population of prey. However, the amount of prey, in turn, also helps to determine the number of predators. Both populations, predators and prey, are strongly linked with each other. The relationship of food chains is a basic mechanism in nature. An interesting way to observe these relationships are food webs. Charles Elton introduced the concept of food webs in his classical book *Animal Ecology* [ELT 27].

The concept of food webs, which is now a basic concept in ecology, tries to represent the relationships, and dependencies, among producers and consumers organizing the elements into functional groups. Groups that have the same predators and prey are considered as functionally equivalent. This organization makes it clear who is higher in the trophic pyramid, as shown in Figure I.1, and helps in the evaluation of how energy, or nutrients, are transmitted from the plants to top predators. In his book, Elton speculates, for example, what the consequences would be of removing wolves from the ecosystem. The result would be the widespread increase of deer, as their natural predators would start to decrease. Interestingly enough, this exact scenario happened and could indeed be verified. In 1915, the US Congress authorized the elimination of the remaining wolves and other large predators from the western states. By the 1930s, they had virtually disappeared from the wild, and effectively the deer population increased vertiginously between 1935 and 1945 [RIP 05].

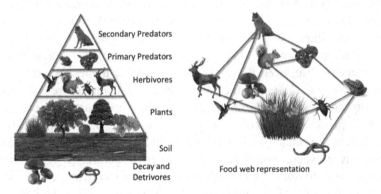

Figure I.1. *Trophic pyramid and a food web representation of the relationship among biological entities. Inspired by Charles Elton's book Animal Ecology [ELT 27]*

I.1. Heuristics and metaheuristics

It is important to note that the original definition of food webs has similarities with the concept of producers/consumers in distributed computing. Thus, we can almost directly use the biological results to control services and elements within computer networks. Some initiatives already exist; for example, Mishra and Ansari [MIS 12] proposed a model based on predator–prey relation to understand the infection of network elements by malicious objects; Gueli [GUE 08] proposed to use the predator–prey relation as a way to model the placement of nodes in a sensor network, just to highlight two propositions.

When we adapt a general concept from one domain and apply that concept to solve problems in another domain, we call this the "concept of metaheuristic". A metaheuristic is a general concept; it is like an umbrella under which solutions to specific problems can be based. A metaheuristic is problem independent, and as such it can be applied to solve a wide range of problems. However, a heuristic, from the Greek *Eὑρίσκω*, which means to find or to discover, is a solution based on the experience that a given procedure reaches a *good enough* result in most cases, even if possibly not reaching the optimal solution. Heuristics exploit problem-dependent information to find a solution to a specific problem. They are used when optimal solutions are too costly, or impossible, to reach with the available time and/or computational resources. We constantly use heuristic approaches in our everyday life, even if we give to them other names. We may call them *rule of thumb*, educated guess, common sense or even stereotyping. All these are heuristics that help us to shortcut complex mental processes to solve a given problem or reach a given conclusion. For example, it is not advisable to jump heads on in a dark water river where you have never jumped before. Why? Simply because you do not have enough information about that specific place and compiling all of it would take a lot of time and effort. That is why you need to measure the deepness of the river at the given area, measure the height of the platform you intend to jump from, calculate the amount of water that would be dislocated with your jump and the deepness you would reach. So we use a shortcut, a rule of thumb or, if you prefer, common sense knowledge that it is not advisable to do so.

We apply these kinds of intuitive rules hundreds of times every day. True, sometimes more successfully than others, but in general they would make sense to most people.

This book mainly discusses heuristic approaches, their inspiration, modeling and application in computer networks field. However, it is important to call attention to the fact that heuristics are neither exact nor fail-proof methods. Even though, in regular cases, it converges to a good result, it is possible that the result of the application of a heuristic presents a result far from optimal. This is of great importance and must be considered when designing a heuristic-based solution.

Another important concept that needs to be clear when using heuristics is the difference between global and local minimum/maximum. In general, when using a heuristic, we are trying to find the maximum or the minimum possible values, e.g. which is the smallest path from node A to node B, or which is the path with the maximum Quality of Service between A and B. "Maximum/minimum local/global" are the terms used to design the desired result when searching for a solution that maximizes/minimizes a given function. Global maximum/minimum is the smallest value a function can have over its entire domain; in general, it is the desired target value. Local minimum/maximum is a value that is smaller/higher than its neighbors, but not necessarily the global one, i.e. the objective of the heuristic is to be as close as possible to a global minimum/maximum, avoiding getting stuck into a local one, and the difference between both may be quite relevant.

The target function can be either a mathematical function, where we have a clear definition of what are the input values and the desired output values, or any other function that we can think of. For example, our function could maximize our chances in a bet over who will win the next Oscar for the best actor. A local maximum would be, among our friends, the one who has the best chances, let's say Fred, because he is the only one who actually took acting classes in high school. A global maximum would be the actual winner, who achieves this by being a professional actor, choosing his plays carefully and working on his career every day, not exactly Fred's case. However, if we were trying to search for the person who has the best chance of winning the next Oscar for best actor, and we start searching for this person close to us, we would first keep Fred in mind, as he is the best candidate so far, i.e. a local maximum. It is fine for a starting point; however, to be effective, the method should converge as fast as possible to candidates who have the maximum chances of being the winner, i.e. the most fitting ones. The minimization problem in this case would be trying to find who, in the world, would have the least chances to win the Oscar. The same line of thinking can be applied to mathematical optimization problems.

I.2. Establish limits

In general, when applying a heuristic, we are interested in solving complex problems, i.e. problems for which it is too difficult, or too costly, to find a perfect solution. Normally, we do not know the answer, sometimes not even the order of magnitude of the best solution. It seems obvious, but if we knew that the answer to the problem, i.e. the target maximization/minimization function, is, let us randomly say, 42 (the answer to the ultimate question of life, the universe and everything – *The Hitchhiker's Guide to the Galaxy*), we would already have the answer; therefore, searching for it would be useless. When searching for something, not knowing the result we want to find, we need to establish limits for the search. Let us say that we are trying to find a perfect gift for Mother's Day. We not only have a budget but also a hard deadline: we need to buy the gift before Mother's Day, otherwise it would be useless. We search for gifts, we establish a list of potential gifts, and in the end, even if we are not fully satisfied with the gifts on our list, we end up buying the best gift on it, as we have a deadline and we cannot search forever. Heuristics work in the same way; we need to define how long we are searching for a solution and when the result is *good enough* for our purposes. This means that in many cases, we will not reach the real global maximum, or we will not know whether the solution we have now is, or is not, the real global maximum solution. It is quite difficult for a heuristic to differentiate a local from a global maximum; they are *best effort* kind of methods. They try their best to find a solution, but in the end there is no guarantee that the solution is the best one. Even though this is certainly a problem in many cases, we may sometimes have no other options.

I.3. Complexity

When studying the computational complexity of algorithms, we can classify them in accordance to difficulty in finding a solution for it. Among the defined classes of algorithms, we have two classes that are specifically interesting: polynomial (P) and non-deterministic polynomial (NP) time. Algorithms that fit into class P are the algorithms that can reach their solution in a deterministic machine using a polynomial amount of computation time or polynomial time. This means that for solving this problem, we have an algorithm that can be implemented on a computer and can find the solution to it with a *reasonable* amount of computation. Sometimes we need to use heuristic techniques for solving problems of this class because the reasonable time may still do not fit our constraints in terms of available time or information available to run the algorithm and find the solution. In these cases, the most common approach is to apply some sort of heuristics to solve the problem.

NP problems are a class of problems that cannot be solved with a deterministic machine in a time that is bounded by a polynomial expression. This means that they cannot be efficiently solved with the available computers. However, if we had a non-deterministic machine, we could find a solution to these problems in polynomial time. In theory, this non-deterministic machine could find the path in the algorithm implementation that leads to the optimal solution in polynomial time.

One of the most fundamental questions in the theory of computation is whether P is equal to NP, i.e. whether we have polynomial solutions for all problems we today classify as NP. It is widely believed that this is not the case, but we have no formal proof of this. Among the subclasses of NP, the subclass that presents a set of particularly interesting problems is called NP-complete. An NP-complete problem is a problem that is NP and we do not have an easy way to solve, i.e. the most difficult problems on NP. The concept was first introduced by Stephen Cook in 1971 [COO 71]. To prove whether a problem is NP-complete or not, we need to be able to solve another NP-complete problem using the solution to the problem for which we want to verify the completeness. In this way if we find a deterministic polynomial time solution to an NP-complete problem, then we would have solutions for all other NP-complete problems if we reduce them to this specific solution. The first problem to be proved NP-complete was the Boolean satisfiability problem (SAT). It consists of determining whether there exists a solution that satisfies a given Boolean formula. In 1972, Richard Karp presented a list of 21 NP-complete problems [KAR 72], which further increased the interest for these kinds of problems. Among the problems in computer networks that are well-known NP-complete, we have the minimum spanning tree, clique, dominating set, graph coloring, longest path and vertex cover.

I.4. Heuristics and nature

Even though they do not guarantee to reach the best possible result, heuristics/metaheuristics are powerful tools to solve a large series of problems. Nature is an incredible source of inspiration. The world is a complex and dynamic environment with an enormous diversity of elements. To overcome the difficulties and challenges of surviving in such a complex and, in some ways, dangerous environment, biological organisms have to evolve, self-organize, self-repair and procreate in order to flourish. All these take into account only local information, and no central control exists. When inserted into a vast and more complex environment, each organism does its best with the information it has locally available. In general, the rules each organism applies are simple and, from time to time, they collaborate with other organisms, even from different species. The new generations of computer networks have been greatly increased in size and complexity. Networks are now starting to face the same kind of challenges nature has been dealing with, and finding solutions to, for millennia. Indeed, it would be interesting if computer networks presented the same efficiency and robustness we find in biological systems[1]. This parallelism between nature and large network structures has not gone unnoticed. The research community is now turning its attention to nature in search of inspiration on how to use bio-inspired methods to solve a large range of problems. For example, we are realizing that centralized structures are not the most efficient and reliable way to control large structures. Bio-inspired strategies have been applied to a large range of problems, but normally these problems present some of these characteristics:

– the absence of a complete and standard mathematical model;

– large number of interdependent variables;

– nonlinear systems;

– combinatorial or extremely large solution space.

Another prominent characteristic of biological systems, which in fact brings them even closer to computer networks, is that, in nature, communication normally plays an important role. In fact, communication is a fundamental part of the organization of most biological systems on different levels of evolution and organization. For example, ants use pheromones to communicate; without this, the whole strategy they use would not be possible. The same could be said about how human neuronal networks are organized. If neurons were prevented from exchanging electrical signals among each other, the whole brain structure would be useless. However, the importance of exchanging information is not only important for small-scale or microscopic systems. Imagine our own society without communication. What would the world be like if we could not exchange experiences, ask for impressions and advices and learn from others' mistakes. One of the greatest differences between human society and other less developed natural social structures lies in the complexity and expressiveness of our communication methods. These and many other natural networks seem to be much better organized,

1 Even though "biological systems" are a subset of "natural systems", in the context of this book we will use these two terms interchangeably. A natural system is the one found in nature, without human intervention. From this description, the gravity, solar and tide systems are natural, but not biological. A biological system requires life to be involved somehow, i.e. "bios" is *life* in Greek.

and work much more efficiently than computer networks. As mentioned, it is only natural for us to search for links and inspiration in biological systems to help solve problems related to artificial networks.

I.5. What to choose

We now know what heuristics and metaheuristics are, and why we sometimes need them; now the problem that arises is how to use all of these to solve our own problems. We have two possible solutions. The first solution is to use a heuristic that is already available in the literature and adapt it to solve the target problem. In this case, we should search for heuristics that can give good answers and could be adaptable to the problem at hand. If by chance someone has already solved the problem using a given heuristic, then we know exactly what the expected results are. In this case, the problem becomes the implementation of strategies and defining the parameters to get the best possible results.

However, if no one has ever applied any heuristic to solve the problem, then we should search over the metaheuristics, the ones that are most likely to find good solutions, and adapt the chosen meta-heuristic(s) to the target problem. Another option is to adapt a heuristic that someone has used in a different context and use it to solve the problem at hand. When adapting a heuristic or a metaheuristic to a new problem, we need to focus on the quality of results reached, as they need to be evaluated very carefully. The task of choosing the best approaches to be used and adapting them to the problem at stake is difficult and undoubtedly important, but the most important part would be to evaluate the performance of such solutions in giving the desired result. We can implement the best heuristic ever created; if it is not well adapted to solve the problem at hand, it will not give good enough results; therefore, there is no point in using it. "If you ain't got no axe, you can't cut no wood! Gentlemen, you got to use the right tool for the job!"– attributed to John Henry Eaton, an American politician. This is true for any kind of work;

the best results are normally achieved when the right tools are applied.

However, even when applying the right tools one needs to know how far he/she is from the perfect or desired result, even if this best result is not achievable with the available resources. This tracking of the reached solutions, compared with the objectives, will not only give us a hint of how good our present set of solutions are, but also give us a hint of when we should stop searching for new solutions. Evaluating the results to find the *"good enough"* solution is part of the job and is a highly relevant one.

The second possibility is to find an inspiration on your own, i.e. to create a heuristic/metaheuristic to solve the target problem. For this, we need to learn, interpret and translate the lessons from nature to the computer networks language. The basic idea is to find a natural system presenting high-level characteristics similar to the ones we wish the computational system to present[2]. From this initial observation, we can explore deeper the natural system to understand its parts and how they interact/cooperate to present the desired behavior/results. Ideally, the components and their behavior should be mathematically and/or algorithmically expressed. The mathematical analysis of the natural system helps us in its understanding and, therefore, its translation from the natural world to the artificial world. First, we should try to imitate the natural behavior as closely as possible to try to find the primary phenomena, i.e. the components of the natural system responsible for the behavior we are interested in. This method of study is called reductionism. This technique is largely used by biologists for analyzing natural systems [BRI 14], even though recently it has raised some criticisms [REG 04].

---

2 We can call these observable characteristics of phenotype (from the Greek *phainein*, which means to show, and *typos*, which means to type). Phenotype is the set of observable characteristics, or traits, of an organism.

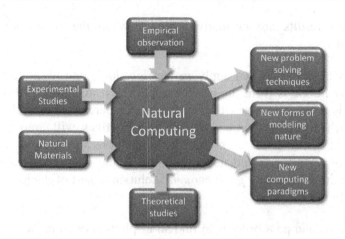

Figure I.2. *Approaches used in the development of natural computing. Inspired by Castro [CAS 07]*

Constituent reductionism or methodological reductionism [WIN 06] believes that when a complex system is divided, the resulting parts are exactly the ones that compose the system; nothing else is left. For example, a living organism can be dissected down to its molecular level. Even if the organism dies in the process, the entire set of molecules that composed the body is present. Causal reductionism claims that the whole system behavior can be implied by the simple sum of the effects of each one of its individual components over the others [JON 13]. When dissecting the organism, we prevent interaction among its components; this is why the organism dies in the process. Reductionism does not imply that larger patterns, or behaviors, will arise from the sum of the parts that do not necessarily present such property; in fact, they ignore any *emergent phenomena,* a key concept for artificial intelligence systems. However, reductionist techniques provide some useful tools to understand the emergent phenomena by analyzing its parts and their interactions. The holistic view of a system, trying to determine the full extension and the macroscopic behavior of a system, also requires the view of how the parts interact and behave. Any theory that ignores the internal behavior of the particles composing a system lacks scientific evidence to explain the macroscopic effects, which have their roots in the microscopic processes [GOU 10].

Figure I.3. *Road map for the design of bio-inspired solutions: the first step is a search in the natural world for systems that present the desired behavior understanding the natural system and its components. The second step is to create a realistic, or as realist as possible, computational model of the biological system. The third step is, in general, the process of simplifying the model to capture its essence, fine-tuning the various parameters to improve the performance of the model in order to perform the desired tasks*

I.6. Complex systems

Biological systems as well as computer networks can be characterized as complex systems. Complex systems are systems in which the behavior of the system is difficult to predict and understand because parameters that influence the systems, the causes and effects are not obviously related. Complex systems focus on trying to analyze questions about parts, whole systems and relationships as a way to predict the behavior of the system. This research field is interested in how parts of a system interact with each other, and with the environment, as a way to explain the system behavior, mainly for systems in which the internal micromechanisms greatly differ from the macrobehavior.

Both biological systems and computer networks present quite distinct characteristics and mechanisms when considered from microscopic and macroscopic points of view. In fact, the very essence of what we are as human beings is an indisputable example. We can describe the chemical reactions that happen in each part of our body; we can explain how ingredients are processed and energy is generated within the Krebs cycle [KRE 87]. On a macro level, people use this energy to perform a series of tasks where the origin of the energy is somehow hidden. Even though we correctly understand how the energy is generated and the energy transfers occur, the gap between the micro and macro levels is enormous. The Krebs cycle is a fundamental part of us as living beings, and we need to understand that in order to understand life, but one can hardly explain the full extension of a person starting from that. What complex systems show us is that both holistic and reductionist views are important. Indeed, the reductionist view gives us the basis to develop more holistic theories and explanations. In general, complex systems tend to investigate through microscopic, agent-based models and afterward its results used to model/justify macroscopic behaviors. Each agent, or particle, has its behavior studied, its evolution through time is determined and then the relation to its macroscopic counterpart is established. This approach tends to improve the predictability of the macroscopic models. Even if it is too simplistic, in essence, we can say that the macroscopic behavior is a generalization/simplification of the impacts of the microscopic interactions over the whole system.

This simplification helps us to hide the complexity of the system and makes it easy to understand its possible use as a source of inspiration. Even if someone argues, we can completely express a complex system starting from the iterations of its most basic particles. This kind of approach may be extremely time-consuming for large systems. Identifying the relevant parameters from large-scale experiments is already an exhaustive task, but emulating a system from its basic interactions may be computationally intractable. Macroscopic models have a lower computational complexity and are less dependent from the point of view of the system's size. They are better adapted to work with large volumes of data and, in general, express the properties we need over computer networks.

I.7. Treating limitations

The study of a natural system with the intention of translating its methods to model computational systems also implies the analysis of the original system limitations. Biological systems may be quite complex, and present a number of subtle iterations and epiphenomena. Epiphenomena are the secondary phenomena that occur along with another central phenomenon. In general, the most successful attempts to model biologically inspired systems are the ones that manage to extract the essence of the method and apply the pertinent principles avoiding being constrained by the limitations imposed by the original biological method [MEI 10]. This means ignoring any non-relevant epiphenomena. The analysis of the interactions among the components of the system, not only the parts that compose it, also requires great attention and is considered a factor of success for most attempts [TIM 06]. In general, we should concentrate on the characteristics that make the system what it is. We should learn with the high-level lessons and principles, and then try to adapt these to solve the problems of artificial systems. Modeling attempts that try to mimic too closely the instruments of natural systems also tend to copy the limitations and undesirable effects of those. It is important to note that biological systems are the result of billions of years, adaptation to an ever-changing environment in a trial-and-error kind of strategy. Nature solutions are adapted not only to solve a great deal of problems, but also to carry the whole history of changes used in the adaptation of the solution to the every changing environment.

Nature's random approach certainly presents amazing solutions to problems; however, not only are these solutions bounded by the real world, but also they may present peculiarities that were inherited from the evolutionary process. Solutions to old problems may be difficult to completely eliminate over the next generations, even if the initial target problem does not exist anymore. Vestigial structures in the human body, e.g. wisdom teeth and the appendix, are examples of how long it can take nature to forget structures that were once crucial for the development of a certain species. Modeling these traits would increase the complexity of the final system and, most of the time, not add much, if anything, to the solution. Recognizing and understanding the higher-level principles responsible for the system phenotype may require a strong understanding of biology and the state of the art in biological research [MEI 10]. An in-depth knowledge of biology is not mandatory to model natural systems, but it certainly helps. Relying on the work of biologists in creating a system-level vision of the observed natural system [NOB 08] helps us to provide the definition of more robust models. This interdisciplinarity is beneficial to both fields; computer scientists can benefit from the experience and vision of biologists to model robust and efficient solutions/protocols, but the contrary is also true. Biologists can benefit from the created models to better understand the implications of the natural systems and networks. This collaboration can help us to push the boundaries of knowledge and improve our knowledge and understanding of the fundamental components, relationships and dynamics of the complex ecosystem in which we participate.

I.8. Modeling biological systems

In a nutshell, we can say that modeling a biological system consists of the identification of the analogies, the understanding of its implications and methods, and the extraction of the characteristics that make it an interesting analogy.

When studying a complex system, we should start defining the set of characteristics that capture the system's dynamics, and model these first. These models should be sufficiently general to represent a wide range of possibilities but have sufficient structure to capture interesting features. The basic mechanisms of biological systems, relevant to network systems, that we should pay special attention to are their sensory, communication and signaling mechanisms [BAL 11].

The way biological structures receive, process and deliver information can be better understood when modeled, for example, as a bow tie architecture [CSE 04]. "Bow tie" is a conceptual way to understand some operational and functional biological structures. The name "bow tie" comes from the shape of the model representation, with inputs, processing and outputs. Bow tie structures can handle a great diversity of inputs, fanning into the knot, process in a more specialized and less diverse core, the knot, and output the results in a large range of diverse outputs, fanning out of the bowtie. Figure I.4 shows a representation of this concept. Bow ties are able to optimally organize flows of mass, energy and signals in a structure that is able to deal with the highly non-uniform stimuli it receives from the environment [CSE 04]. These structures present a good trade-off among efficiency, robustness and evolvability. The tight and specialized core is able to concentrate the processing of diverse inputs to produce a large range of useful responses, to possibly different contexts and structures. These kinds of structures are largely spread and observed in different structural organizations of organisms in different levels of evolution.

Even more interesting is the fact that this kind of organization is so natural, or intuitive, that we can even find it in computer networks. Indeed, a router is a knot that receives packets from multiple inputs and decides their destination over multiple outputs. Bow tie architectures help improve the efficiency and self-organization capabilities of biological systems; their identification and understanding may lead to a better comprehension of how the natural system structures behave to perform their roles.

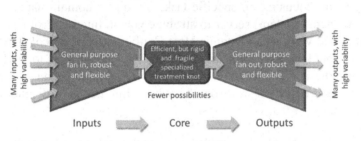

Figure I.4. *Schematic representation of a general bow tie architecture. A tight and specialized knot treats different inputs and produces a wide range of outputs. The inputs are somehow transformed to be adapted as an input that the knot knows how to handle. In general, the knot is quite efficient in doing the task that it is supposed to do; however, it is not robust treating, for example unexpected values. Inspired by Tieri et al. [TIE 10]*

Natural systems were optimized over thousands of years and the power laws involved in these systems make them scalable, reliable and robust. A power law is a relationship between two elements where the quantity of one element is proportional to a fixed power of the second element. If the quantity of one element varies, the quantity of the second element varies as a function of the power of the quantity of the first element. This kind of relationship is important because it reveals an underlying regularity in the properties of systems. Not all natural systems present such characteristics, but a majority of them do; for example, the frequency of earthquakes varies inversely to their intensity. Carlson and Doyle [CAR 99] called such systems highly optimized tolerance (HOT) systems. For Carlson and Doyle, the power laws present in some natural systems are the result of trade-offs between cost of resources and tolerance to risks. The key assumption for HOT systems is that a complex internal structure is required for systems to exhibit robust external behavior. It is a fundamental trade-off between structural simplicity and robustness. For example, living beings, in general, have a series of strictly organized internal structures that work together to maintain life. Each one of these organs has a specific function and, in general, they are not interchangeable; for example, a liver cannot take the place of the heart for pumping blood through the body. The same applies to networks; we have a complex internal structure, with specialized components focusing on specific tasks. Asking a domain name DNS (domain name system) server to attribute you an Internet Protocol (IP) number and work as a Dynamic Host Configuration Protocol (DHCP) server would be as useful as expecting the human skin to perform photosynthesis. The domain name service was not conceived to provide addresses; its role on the whole Internet is a distinct and precise one.

The HOT theory also implies that highly optimized systems are only robust to expected failure scenarios, and can become quite fragile in the presence of unknown, or unexpected, situations. For example, we are prepared to ingest a series of foods, break them into smaller and simpler elements and afterward digest them. This is a regular behavior, and we perform it quite well and efficiently. However, our whole system may be perturbed by small quantities of substances not commonly found in nature. For example, ethyl alcohol (C_2H_6O) is not common in our regular natural food sources, and a relatively small ingestion of it may encourage extremely non-rational behavior. Just for the sake of example, we may be convinced that calling an ex-girlfriend at 4 am is a really good idea, and by that time, even be convinced that it is indeed the right thing to do. Similarly, in computer networks, simple misconfigurations or single faulty components may have great consequences. In 2013, a large number of users of the state of New York, in the USA, were offline for 24 h due to a failure in a single fiber-optic switch. Also in 2013, a connection problem between the two biggest operators of the US stock exchanges brought half of Nasdaq network down. Nasdaq is one of the largest stock markets in the world; thus illustrating the possible consequences of a simple failure in computer network systems.

The characteristic features of HOT systems include [CAR 99]:

– high efficiency, performance and robustness to designed-for uncertainties;

– hypersensitivity to design flaws and unanticipated perturbations;

– non-generic, specialized and structured configurations;

– power laws distributions.

Natural systems are generally complex dynamic systems, and many researchers have dedicated their lives to understanding them. We already have a large set of theories and models (e.g. bow tie and HOT) in the literature, created to try to understand and analyze such systems. It can be quite advisable to get acquainted with the way researchers have modeled complex systems with the mathematical and statistical apparatus created to understand these systems. For example, knowledge of chaos theory, control theory, ergodic theory, functional analysis and graph theory can ease the comprehension of complex systems and will provide a mathematical and theoretical background to help describe the target system in a formal and precise manner.

I.9. Classification of biological systems

Sipper *et al.* [SIP 97] proposed a three-dimensional (3D) classification framework for describing bio-inspired systems. The classification method is based on three levels of organization, and is called the phylogeny, ontogeny and epigenesis (POE) model. It was initially proposed to classify bio-inspired hardware systems, where the nature system is ranked according to its influences over three orthogonal axes: phylogeny (P), ontogeny (O) and epigenesis (E).

The term "phylogeny", derived from the Greek φυλή (phylé), meaning tribe, clan or race, and γένεσις (genesis), meaning origin, makes a reference to the evolutionary history of the species and the relationship among groups of such organisms. It is related to the reproduction/multiplication of elements of one species and the way the genetic material is transferred over generations. On an individual level, the reproductive process presents a low error rate; mutations are a rare event. However, from the species point of view, mutations help improve the diversity of the population, which is fundamental for the survival of a species due to the ever-changing characteristic of nature. Phylogenetic mechanisms are non-deterministic, where mutation and recombination provide the biggest sources of diversity.

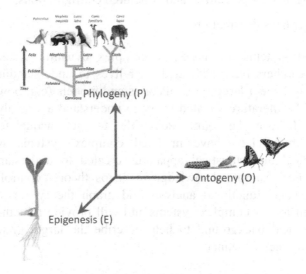

Figure I.5. *The POE model that divides the space of bio-inspired systems into three axes: phylogeny, ontogeny and epigenesis. Adapted from [SIP 97]*

The term "ontogeny" is derived from the Greek ὄντος (ontos), meaning to be, and γένεσις (genesis), meaning origin. It is related to the developmental history of an organism within its own lifetime, i.e. the development of the individual organism. In a nutshell, individuals develop (*ontogeny*), while species evolve (*phylogeny*). An example of ontogeny is the successive divisions of initial cell generating specialized tissues and organs. In practice, ontogeny refers to the development process of multicellular organisms and is a deterministic process. An error in one single phase may lead to unpredictable and, typically, fatal consequences.

The term "epigenesis" is derived from the Greek ἐπί (epi), meaning over or around, and γένεσις (genesis), meaning origin. Epigenesis studies in the inheritable changes that are not caused by changes in the deoxyribonucleic acid (DNA) sequence. It also refers to the study of stable, long-term alterations in the transcriptional potential of a cell that are not necessarily heritable. In general, changes occur within the lifespan of one individual; however, if the change occurs in a sperm or egg cell that is supposed to be fertilized, then some epigenetic changes may be transferred to the next generation [CHA 07]. More commonly, epigenesis refers to the adaptation capacity of an individual as a response to changes over its environment. An example is the learning process. Individuals are born with a set of systems defined in their genome (e.g. nervous, endocrine and immune system) which suffer modifications during their lifetime to evolve and adapt to different stimuli of the environment. The brain changes constantly, and it is precisely because of such changes that we are able to learn. During the time you are reading this paragraph, at this very moment, your brain is changing. If, some time in the future, you remember this single fact, mention it to someone, for example. This is proof that your brain suffered a long-term change while you were reading this specific paragraph. We can also say that traditions and the transfer of knowledge from one individual to another are also parts of an epigenic process. Oral traditions, family, books, schools and many other things take part in the epigenic process of each individual and, thus, perpetuate epigenesis over generations.

Generally, we can understand the three axes proposed in [SIP 97] as epigenesis being a learning process, ontogenesis as being the development of a single individual, regarding his/her own genetic material, without external influences and phylogeny as being the species evolutionary process. We can use this classification to better understand the bio-inspired methods we are willing to study and classify the heuristics over the axis. Figure I.5 presents the three orthogonal axes. Neural networks and machine learning techniques are classified on the epigenesis axis. Genetic algorithms and evolutionary techniques are classified on the phylogeny axis. Works on cellular automata and their self-reproducibility aspects can be classified on the ontogeny axis. Some researchers also argue that the simple classification on the axis is not enough; so they define planes on to which the methods can be fitted. For example, Miorandi *et al.*

expressed the PO plane where they classify techniques that present phylogeny and ontogeny characteristics (e.g. chemical genetic programming [PIA 04], emergent self-repair [AND 06], evolutionary morphogenesis [ROG 07] and genetic regulatory networks [BON 02]).

I.10. Self-organization

One of the most important reasons that bio-inspired systems interest us is related to their self-organizing characteristics. In natural systems, organization dynamics arise spontaneously from their internal structure. Self-organizing natural systems are typically nonlinear and unpredictable [ZHE 12]. The feedback interactions between the parts of the system are constantly searching for a stable organization point. Nonlinear systems have, in general, different attractors. An attractor is a stable point; it may be a fixed or limit cycle one, which leads to a non-chaotic state or unstable equilibrium state. Positive feedback increases the system dynamics, which is absorbed and transformed to reach a stable configuration, which is an attractor.

Natural systems, in general, have a tendency to converge to a stable and organized state. Even in the presence of perturbations, the system tends to return to a stable point because of the attractors. Strangely enough, in general, the bigger the perturbation, the faster the system tends to auto-organize. The explanation for this is that the more perturbed a system is, the higher the probability of finding an attractor. No self-organization is possible if the system is not perturbed and the attractor is reached. We say a system is in fluctuation when it is caught between actractors. Fluctuations are random deviations of a system from its average state, which pushes it into either one of its attractors [PRI 84]. Entropy is the number of ways a system can be organized. It is used to measure the state of organization of a system. Entropy can be understood as the measure of disorder of a system. The second law of thermodynamics states that entropy of an isolated system never decreases. The given system will naturally evolve toward an equilibrium point, the configuration with maximum entropy. Systems that are not isolated may decrease in entropy if they increase the neighbor systems in entropy of the same amount. Living organisms show dissipative spontaneous dynamics. Plants and animals take in energy and matter in a low entropy form as light or food and export it back in a high entropy form, as waste products. This allows them to reduce their internal entropy, thus counteracting the degradation implied by the second law of thermodynamics.

However, for large systems to maintain or increase their self-organization, the system needs to export more entropy than is produced by its internal metabolism [DEN 51]. In thermodynamics, the *exergy* (from the Greek *ex* and *ergon*, from work) is the maximum useful work performed by a system during a stabilization process in which the system goes to an equilibrium state. Günther [GUN 94] argues that the entropy is produced by a biological system in the form of low-exergy products, e.g. water vapor, carbon dioxide and inorganic ions. The environment then uses these in a regenerative cycle so that there is no accumulation of low- exergy products, i.e. "garbage". In some sense, biological systems are interconnected since we produce the raw materials that afterward will be used by others. When studying natural systems, it is often useful to remember Antoine Lavoisier's conservation principle "In nature nothing is created, nothing is lost, everything changes".

In the chapters of this book, we will discuss a series of biologically inspired methods and their applications in the context of computer networks. For each method, we will present their biological roots, how the system works in the natural world, and after that how some researchers understand the methods of the biological system and adapt them to solve a series of different problems in the context of computer networks.

I.11. Bibliography

[AND 06] ANDERSEN T., NEWMAN R., OTTER T., Development of virtual embryos with emergent self-repair, Technical Report FS-06-03, Developmental Systems, AAAI Fall Symposium, 2006.

[BAL 11] BALASUBRAMANIAM S., LEIBNITZ K., LIO P. *et al.*, "Biological principles for future internet architecture design", *IEEE Communications Magazine*, vol. 49, no. 7, pp. 44–52, July 2011.

[BON 02] BONGARD J.C., "Evolving modular genetic regulatory networks", *IEEE 2002 Congress on Evolutionary Computation (CEC '02)*, vol. 2, pp. 1872–1877, 2002.

[BRI 14] BRIGANDT I., LOVE A., "Reductionism in biology", in ZALTA E.N. (ed.), *The Stanford Encyclopedia of Philosophy*, available at http://plato.stanford.edu/archives/win2014/entries/reduction-biology/, Winter 2014.

[CAR 99] CARLSON J.M., DOYLE J., "Highly optimized tolerance: a mechanism for power laws in designed systems", *Physical Review E, Statistical Physics, Plasmas, Fluids, and Related Interdisciplinary Topics*, vol. 60, no. 2, pp. 1412–1427, 1999.

[CAS 07] CASTRO L.N., "Fundamentals of natural computing: an overview", *Physics of Life Reviews*, vol. 4, pp. 1–36, 2007.

[CHA 07] CHANDLER V.L., "Paramutation: from maize to mice", *Cell*, vol. 128, no. 4, pp. 641–645, February 2007.

[COO 71] COOK S.A., "The complexity of theorem proving procedures", *Proceedings of the 3rd Annual ACM Symposium on the Theory of Computing*, ACM, New York, pp. 151–158, 1971.

[CSE 04] CSETE M., DOYLE J., "Bow ties, metabolism and disease", *Trends in Biotechnology*, vol. 22, no. 9 , pp. 446–450, 2004.

[DEN 51] DENBIGH K.G., *The Thermodynamics of the Steady State*, Methuen & Co., Ltd., 1951.

[ELT 27] ELTON C., *Animal Ecology*, Macmillan Co., New York, 1927.

[GOR 00] GORSHKOV V., GORSHKOV V.V., MAKARIEVA A.M., *Biotic Regulation of the Environment*, Springer-Verlag, London, 2000.

[GOU 10] GOULD H., TOBOCHNIK J., *Statistical and Thermal Physics: With Computer Applications*, Princeton University Press, 2010.

[GUE 06] GUELI R., "Predator – prey model for discrete sensor placement", *8th Annual Water Distribution System Analysis Symposium*, Cincinnati, OH, August 27-30, 2006.

[GÜN 94] GÜNTHER F., Self-organisation in systems far from thermodynamic equilibrium: some clues to the structure and function of biological systems, PhD Thesis, Department of Systems Ecology, Natural Resource Management Institute, Stockholm University, Sweden, June 1994.

[JON 13] JONES R.H., *Analysis & the Fullness of Reality: An Introduction to Reductionism & Emergence*, CreateSpace Independent Publishing Platform, 2013.

[KAR 72] KARP R.M., "Reducibility among combinatorial problems", *Complexity of Computer Computations*, Plenum, New York, pp. 85–103, 1972.

[KRE 87] KREBS H.A., WEITZMAN P.D.J., *Krebs' Citric Acid Cycle: Half a Century and Still Turning*, Biochemical Society, London, 1987.

[MEI 10] MEISEL M., PAPPAS V., ZHANG L., "A taxonomy of biologically inspired research in computer networking", *Computer Networks*, vol. 54, no. 6, pp. 901–916, April 2010.

[MIO 10] MIORANDI D., YAMAMOTO L., PELLEGRINI F., "A survey of evolutionary and embryogenic approaches to autonomic networking", *Computer Networks*, vol. 54, no. 6, pp. 944–959, April 2010.

[MIS 12] MISHRA B.K., ANSARI G.M., "Predator-prey models for the attack of malicious objects in computer network", *Journal of Engineering and Applied Sciences*, vol. 4, no. 3, pp. 215–220, 2012.

[NEW 97] NEWMAN M., "A model of mass extinction", *Journal of Theoretical Biology*, vol. 189, pp. 235–252, 1997.

[NOB 08] NOBLE D., *The Music of Life: Biology Beyond Genes*, Oxford University Press, 2008.

[PIA 04] PIASECZNY W., SUZUKI H., SAWAI H., "Chemical genetic programming, the effect of evolving amino acids", *Genetic and Evolutionary Computation Conference 2004 (GECCO '04)*, 2004.

[PRI 84] PRIGOGINE I., STRENGERS I., *Order Out of Chaos*, Bantam Books, New York, NY, 1984.

[QUA 08] QUALCOM, Interferometric modulator (IMOD) technology overview, Qualcom white paper, May 2008.

[REG 04] REGENMORTEL M.H.V.V., "Reductionism and complexity in molecular biology", *EMBO reports*, vol. 5, no. 11, pp. 1016–1020, November 2004.

[RIP 05] RIPPLE W.J., BESCHTA R.L., "Linking wolves and plants: Aldo Leopold on trophic cascades", *BioScience*, vol. 55, pp. 613–621, 2005.

[ROG 07] ROGGEN D., FEDERICI D., FLOREANO D., "Evolutionary morphogenesis for multi-cellular systems", *Genetic Programming and Evolvable Machines*, vol. 8, no. 1, pp. 61–96, 2007.

[RYA 97] RYAN J., LIN M.J., MIIKKULAINEN R., Intrusion detection with neural networks, AAAI Technical Report WS-97-07, 1997.

[SAL 96] SALEH K., "Synthesis of communications protocols: an annotated bibliography", *SIGCOMM Computer Communications Review*, vol. 26, no. 5, pp. 40–59, October 1996.

[SIP 97] SIPPER M., SANCHEZ E., MANGE D. *et al.*, "Phylogenetic, ontogenetic, and epigenetic view of bio-inspired hardware systems", *IEEE Transactions on Evolutionary Computation*, vol. 1, no. 1, pp. 83–97, April 1997.

[TIE 10] TIERI P., GRIGNOLIO A., ZAIKIN A. *et al.*, "Network, degeneracy and bow tie. Integrating paradigms and architectures to grasp the complexity of the immune system", *Theoretical Biology and Medical Modelling*, vol. 7, no. 32, p. 32, 2010.

[TIM 06] TIMMIS J., AMOS M., BANZHAF W. *et al.*, "Going back to our roots: second generation biocomputing", *International Journal of Unconventional Computing*, vol. 2, no. 4, pp. 349–378, 2006.

[TIV 12] TIVEL D., *Evolution: The Universe, Life, Cultures, Ethnicity, Religion, Science, and Technology*, Dorrance Publishing, 2012.

[VOG 88] VOGEL S., *Life's Devices: The Physical World of Animals and Plants*, Princeton University Press, 1988.

[WAL 07] WALDROP M.M., "Brilliant displays: a new technology that mimics the way nature gives bright color to butterfly wings can make cell phone displays clearly legible, even in the sun's glare", *Scientific American*, vol. 297 no. 5, pp. 94–97, November 2007.

[WIN 06] WINTHER R.G., "Parts and theories in compositional biology", *Biology and Philosophy*, vol. 21, pp. 471–499, 2006.

[ZHE 12] ZHEGUNOV G., *The Dual Nature of Life*, The Frontiers Collection, Springer, 2012.

1

Evolution and Evolutionary Algorithms

*Word cloud representing the full text of this chapter and the
words frequencies. Created with Wordle.net.*

Evolutionary algorithms are the algorithms that are based on the evolution of the species; in general, they are based on the main evolutionary theory of Charles Darwin [DAR 59]. The way the evolutionary mechanisms are implemented varies considerably; however, the basic idea behind all these variations is similar. Evolutionary algorithms are characterized by the existence of a population of individuals exposed to environmental pressure, which leads to natural selection, i.e. the survival of the fittest, and in turn the increase of the average fitness of the population. Fitness is the measure of the degree of adaptation of an organism to its environment; the bigger the fitness is, the more the organism is fit and adapted to the environment. In general, evolutionary algorithms focus only on a subset of mechanisms defined over the biological evolutionary process.

The main natural methods that influence artificial evolution are reproduction, mutation, recombination, natural selection and survival of the fittest. In general, the problem is coded as individuals in a population and a cost function plays the role of the environment, defining which of the individuals in the population are the fittest. The evolution of the population then takes place with successive applications of the evolutionary mechanisms. The main operators of the recombination process are recombination and mutation. They create the required diversity in the population while the selection, at each generation, brings the population increasingly closer to the solution.

Evolution is a constant process in nature and implies, over successive generations, the change in the inherited characteristics of the population. This change occurs naturally during the recombination process, as well as through mutation. This efficient method to increase the diversity in the population is applicable to each living species, from the smallest bacteria to the giant sequoias.

1.1. Brief introduction to evolution

The evolution theory tells us that all life on the Earth descended from one universal single cell ancestor that lived around 3.5–3.8 billion years ago [DOO 00]. The level of biodiversity that we have today is the result of the repetitive application of the reproductive operators for thousands of millions of years over the descendants of this common ancestor. Traits of the repeated process are printed in the parts of the genetic material that all living beings share. The name of the field that tries to track and analyze these hereditary differences in the deoxyribonucleic acid (DNA) sequences is called molecular phylogenetics. This field of research started in the mid-1960s with the work of Emile Zuckerkandl and Linus Pauling [ZUK 65] who came out with this revolutionary idea for the first time. Instead of only looking at anatomy, or physiology, traits to understand the relations between living organisms, Zuckerkandl and Pauling based their analysis on the chemical differences in the organization of the building blocks of genes and proteins. This bottom-up approach revolutionized our understanding of the world and the relations among living beings. Species that share a more recent ancestor have larger amounts of genetic material in common than those who have a more ancient ancestor. This can be used to reconstruct and tell the history of species evolution and also of their extinction.

Even though the work of Zuckerkandl and Pauling exposed the evolutionary methods in a way that had never been done before, the first to formulate a valid scientific argument in favor of the theory of evolution, by means of natural selection, was Charles Darwin [DAR 59]. The main arguments that led Darwin to develop his theory came mainly from three facts observed on diverse general populations. First, more offspring are produced than could possibly survive. Second, characteristics vary among individuals of the population, which leads to different rates of survival and reproduction. Finally, the third fact is that some of these characteristics are heritable, i.e. they survive between generations on the same reproductive line of individuals. This repetitive process creates and preserves traits that improve the chances of survival of individuals; the most suited traits for the specific functional roles they play tend to be preserved. The world is competitive and any trait that gives some sort of advantage to an individual over others tends to be preserved. Moreover, as the advantageous traits tend to be preserved, the tendency is that the population improves over the generations. The next is more adapted to survive and reproduce over the present environment than the previous ones.

Evolutive methods work by changing heritable traits of a particular organism. This led to the work of another important researcher, Gregor Mendel. Mendel is considered as the founder of genetics as he was the first to try to scientifically understand inheritance as a trait that lasts over generations. Mendel's seminal paper "Experiments in Plant Hybridization" [MEN 66], in 1866, is the first paper which collected scientific evidences and clearly proved that each organism holds a set of physical traits, corresponding to invisible information, coded within each individual. Moreover, these invisible information elements, now called genes, exist in pairs. Mendel showed that only one member of this genetic pair is passed onto each progeny. At that time, Mendel had no information about chromosomes, cell structure, or the fertilization or the mitosis and meiosis processes. His study was more of an amazing work of statistical observation and deduction much ahead of his time. Today, his work retains a strong influence in the field of genetics.

In his research, Mendel studied the impact of crossbreeding peas and how their physical characteristics were carried over generations. Peas are easy to cultivate and present a number of distinctive traits which make them a good study case. Mendel questioned how physical characteristics of peas were transmitted from generation to generation and whether these characteristics were modified over time by the influence of the seed's genetic material. The observed and counted phenotypical traits were:

1) shape of the seed – round or wrinkled;

2) color of the pea – yellow or green;

3) color of the seed – gray or white;

4) form of the ripe pod – inflated or constricted between peas;

5) color of the unripe pod – green or yellow;

6) position of the flower – terminal or axial;

7) length of the stem – tall or short.

Mendel's work was meticulous and presented the proportions of each one of these traits within a series of distinct peas crossbreeding experiments. It is clearly stated that some characteristics would be dominant over others, and it was also observed that some traits even if not present in one generation would reemerge over subsequent generations. The observed proportion on dominant/recessive traits over a generation was 3:1. The series of genes responsible for a specific trait is called allele.

Mendel's law of segregation states that each member of a pair of alleles maintains its own integrity, regardless of which one is the dominant. At reproduction, only one allele of a pair is transmitted to each gamete, and that choice is entirely random. Moreover, Mendel also observed that each of the traits he was following sorted themselves independently. Mendel's law of independent assortment states that characteristics that are controlled by different genes will manifest independently from all others. For example, the color of the peas and their shape traits did not influence each other.

The work of Mendel proves that the evolution in organisms occurs through changes in heritable traits, i.e. the particular characteristics of an organism. In humans, for example, eye color is an inherited characteristic and an individual might inherit the "brown-eye trait" from one of their parents [EIB 96]. Inherited traits are controlled by genes and the complete set of genes within an organism's genome is called its genotype. To simplify, and to make observations more precise, nowadays geneticists distinguish between genes and their expression.

A genotype is the actual set of genes that an organism has. It is the real expression of its genetic material and it is unique to an individual. A phenotype is a measurable characteristic of an organism, the expression of a gene, e.g. eye color, hair color, hair growth, the shape of the nose, number of arms or even biochemical traits such as cholesterol levels, hormones and blood type. Even though, in general, phenotype is determined by the genetic material of each individual, some characteristics are the result of the genetic material plus the interaction of the individuals with the environment they are inserted into. For example, depending on the diet, the cholesterol levels of a person may change drastically during his/her lifetime. Another more concrete example is ability to tan. It is determined by the interaction of genetic material, i.e. the individual predisposition to get tanned and the sunlight. The specific sun-tan that a given individual has is not an inherited trait, since it depends a lot on the individual's exposure to the sun. For the individual's phenotype traits, both components are important: the genetic predisposition and the environment into which the individual is inserted. For populations, the genetic material differentiation accounts for the greatest part of the differentiation among individuals [PEA 06]. Variations in a population's genetic material may be the result of mutation, sexual reproduction or migration of individuals among populations.

1.2. Mechanisms of evolution

1.2.1. *DNA code*

DNA is the molecule that stores the genetic information for growth, division and function of living beings. It is composed of two large sequences of nucleotides twisted into a double helix, joined by hydrogen bounds, as depicted in Figure 1.1. The basic elements that form the DNA that codifies the genetic information are adenine, thymine, cytosine and guanine. These nitrogenous bases pair as follows: adenine pairs with thymine and guanine pairs with cytosine.

In general, in artificial genetic algorithms (GAs), we could say that the DNA is our genetic pool, i.e. the set of codified information that we are evolving over each generation step.

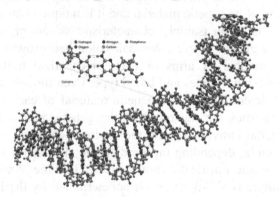

Figure 1.1. *DNA double helix structure, visualized by University of Illinois at Urbana Champaign Visual Molecular Dynamics (VMD) program*

1.2.2. *Mutation*

Mutation is a permanent change in the DNA sequence of an organism's genome that was not present in the genes of its parents. It can be an alteration in the codification of DNA's gene, or a change in the physical organization of a chromosome. Mutations are relatively rare events and are the result of a mistake in the translation of the parent genetic material into its offspring. Even though during the replication of DNA the translation mechanisms are relatively resilient to errors, the number of operations involved is so large that errors may occur. Such errors may or may not have an impact on the characteristics or traits found in the offspring. It all depends on the place where the mutation occurs and the severity of the mutation. The majority of the mutations (~70%) are harmful to the individual. The remaining ~30% are either neutral or slightly beneficial [SAW 07, SAN 04]. Either way, they remain an important factor in the evolutive process. Mutations help improve the variability in the populations' genetic material, which may lead to better and more suited individuals to a specific environment.

The causes of mutations may be exogenous or endogenous factors. Exogenous factors are environmental factors, such as sunlight, radiation and even smoking. Endogenous factors are errors during the DNA replication that can lead to genetic changes.

1.2.3. *Sexual reproduction and recombination*

In sexual organisms, although different between genders, the reproduction process occurs through the recombination of the genetic material of the parents: the offspring has traits from both parents. In asexual organisms, the offspring receives the genetic material from one single parent, and, in general, it is an exact copy of the parent's genetic material. For sexual reproduction, the offspring genome is a random mix of the genes of the parents. Thus, in terms of diversity, organisms that reproduce sexually have a clear advantage over organisms that reproduce asexually. For sexually reproductive individuals, novel combinations of genes are possible in each generation.

Recombination is the name of the process in which the genes of the parents are recombined, i.e. exchanged, to create new and different offspring. The resulting chromosomes are a mix of the chromosomes of the parents and thus distinct from both. Recombination does not alter allele frequencies, but instead changes which alleles are associated with each other, producing offspring with new combinations of alleles. The recombination may occur through either mitosis or meiosis cell division processes.

Meiosis is the process of division of the genetic material of a cell into halves to prepare to pair with the genetic material of the other parent. In meiosis, the DNA replication produces four daughter cells with half the number of chromosomes as the original parent cell. Recombination in this case occurs by pairing the homologous chromosomes of both parents. Mitosis, on the other hand, generates two identical daughter cells that are genetically identical to the parent cell, i.e. it duplicates the DNA and the two new daughter cells that have the same pieces and genetic code. The interchange of sections between homologous chromosomes, the crossover process, supposedly occurs in both mitosis and meiosis. The difference is that when the chromosomes are distinct, as in the case of meiosis, it has a much bigger impact. When the aligned chromosomes are identical, just copies of the same from the parent, the crossover swap results again in identical chromosomes.

1.2.4. *Natural selection*

Natural selection is the process in which nature favors the most adapted organisms. It guarantees that mutations which provide reproductive or survival advantages to an organism will be preserved and progressively become more common in the future generations of the population. Thus, natural selection is the process that makes advantageous mutations, even if relatively rare in the beginning, to become widespread among the population.

The intuition behind the natural selection concept is quite simple. Consider that mutations exist and are inherent to the process of evolution. Organisms for which a given mutation provides a reproductive, or living advantage, will create more offspring during their lifetime. The reason is that either they are more efficient in the production of the offspring or they will live longer and will, for this reason, have more opportunities to procreate. Either way, their genetic material will be more available over the next generation. If the mutation is inherited by the offspring, the process is repeated over the successive generations spreading even more the advantageous mutation, which at some point may become the rule.

1.2.5. *Genetic drift*

Genetic drift defines the changes in the genetic composition of a population due to a random external event during the sampling of the alleles. Genetic drift is not the same as the natural selection because it is not the result of a better adaptation to the environment, even though it influences the allele frequencies over time.

Genetic drift is an important evolutive factor for small or close populations, and is not relevant for large populations, where natural selection is the predominant process. In small populations, just by chance, the frequency of alleles tends to "drift" up and down over the population, not because they present an advantage but because some other factors happened. It may happen that this drift eliminates other alleles completely from the population, just by chance. For example, imagine that at some point a mutation gave a great advantage to one individual. Unluckily, this individual crosses a predator quite early in his lifetime. Despite the advantage, the beneficial mutation will be lost.

Over time, the changes in alleles may lead to different populations, not as a result of adaptation to the environment but due to the lack of natural selection forces.

1.3. Artificial evolution

In general, artificial evolution methods try to emulate the natural evolution process. They define a population and over many generations evolve this population applying the basic evolutive operators. The objective is to use natural mechanisms to try to find solutions for computationally complex problems. In the 1950s, Alan Turing [TUR 50] proposed that the development of intelligent machines should follow the path of evolution. Turing even defined in his argumentation the process of reproduction, mutation and natural selection. However, Holland in the 1970s was the first to really develop a practical evolutionary method, i.e. genetic algorithms [HOL 92]. The description presented here refers to the basic GA mechanisms; however, many other variants exist. It is also worth mentioning that the field of evolutionary algorithms is still an active and evolving research area. A number of new methods and adaptations are proposed every year.

1.3.1. *The basic process*

The main evolutionary principles used in evolutionary methods, in particular GAs, are mutation, reproduction and natural selection. A large part of the work on GA is focused on the adaptation of these concepts over the characteristics of whatever problem we are trying to solve. The first step is to identify the main parameters of the problem and codify them as a gene so that we can apply the evolutionary operators over an initially randomly generated genetic pool afterward. For example, let us assume a communication algorithm that presents a series of configuration parameters, such as the size of the message, time to live and congestion window size, and the best combination of values is unknown as the number of combinations is too large. We could codify each parameter as a gene and use a GA to automatically search for the best combination of parameters. However, in order to determine which genes represent the best combination of values, we need to evaluate their impact on the algorithm. It is for this reason that we need to define a fitness function. The role of the fitness function is to decide which of the individuals are the best candidates, most fit, to solve the given problem.

The basic execution steps of a GA are initialization, evaluation, crossover, mutation and repetition which are explained below:

– *initialization*: this step creates an initial population. In general, this initial population is randomly generated and can be of any size, from a few individuals to thousands;

– *evaluation*: each member of the population is evaluated according to a fitness function. Evaluation measures how fit the individual is in fulfilling the problem requirements, i.e. solving the target problem;

– *selection*: ideally, we want our population's overall fitness to increase, and selection helps us to do so by choosing the best-suited individuals and discarding those who are not well adapted to solve the target problem. There are various ways to perform the selection, but the main idea is to increase the chances for fitter individuals to be preserved over the next generation;

– *reproduction*: during the production or crossover phase, new individuals are created, taking into account the combination of the genetic material of the selected individuals. By randomly selecting two individuals from the pool and exchanging parts of their genes, crossover mimics the natural sex reproduction mechanism. The intention is to create a larger offspring pool of greater fitness. The intuition behind reproduction is that by combining certain traits from two already fit individuals, the offspring will be even "fitter" since it will possibly inherit the best traits from each of its parents;

– *mutation*: even though relatively small, mutation plays a major role in changing the genetic pool. Without mutation, all the combinations that we would ever possibly reach during the successive generations would be already in the initial pool. Mutation typically works by making small changes at random to an individual's genome. In general, it affects a small portion of the population. If the mutation provides an advantage to the individual, it will increase its fitness and the mutation will be carried out by the selection mechanism. If it is not the case, selection will be in charge of removing the mutated gene from the pool;

– *repetitive generation cycle*: after the mutation phase, the generation of genetic pool is complete. Now we can evaluate the pool again and restart the same process for the next generation. The process goes on until we reach the termination condition;

– *termination condition*: there are many reasons to end a GA. The first reason is that the process found a very good solution. However, it could be that the algorithm stopped to converge, e.g. in the last N interactions, and the fitness did not increase more than a predefined threshold. Other reasons for termination could be constraints such as available time and money.

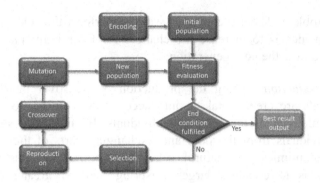

Figure 1.2. *General scheme flowchart of the phases of an evolutionary algorithm*

1.3.2. *Limitations*

Evolutionary algorithms are powerful, but they have limitations and problems like any other methods. The main problem associated with them is to decide whether the found solution is a local or a global maximum/minimum. When applying GA over a problem, normally we go increasingly closer to a solution. However, the problem is how to decide whether this solution is the overall best or just a local one.

GA also loses much of its power if the genetic pool loses its diversity. If the fitness function is not smart enough to evaluate different aspects of the solution, or the mutation rate is not tuned correctly, the genetic pool may be locked around a local maximum/minimum. This premature convergence can lead to a loss of diversity in the genetic pool (genetic drift); given that the best-suited individuals are always used to create the next generations, at some point all the individuals in the population may start to become similar to the local best solution. It is very difficult for GAs to decide whether they got stuck near a local best or whether this is really the global best solution. It is in this kind of scenario that mutation plays a major role, as it can diversify the population. Some works even propose that if a genetic drift is detected, the GA could enter into a *hyper-mutation* mode. In this state, for a limited period of time, the GA mutation rate greatly increases to try to get out from a possible local minimum. Another option is to use an island-based model to evolve the population [WHI 98], i.e. we can divide the genetic population into subpools that evolve separately and in parallel. A small number of elements can migrate from one pool to another; this process increases the diversity of the overall population.

Another problem of GAs is precisely the definition of a good fitness function. The fitness function is the element that is responsible for driving the GA toward the best solution. If the fitness function does a lousy job guiding the pool toward a solution, the whole process is compromised. Unfortunately, for real-world problems, fitness functions that take into account different aspects of the problem may be fairly complex. If we take into account that the function needs to be applied to each individual on the pool, for each generation in the process, GA may become computationally intractable.

Furthermore, another characteristic that may be a problem in some scenarios is that the better solution is always a comparative measure. We compare the best of this generation with the best of the previous generation, which does not necessarily help in determining the real global best.

1.4. Applications on networks

1.4.1. *Network positioning*

The design of network topologies is a complex problem; in fact, it is proved to be a non-deterministic polynomial-time hard (NP-hard) [CLE 99]. The problem gets even more complex when observed from the point of view of networks with hard energy constraints, such as wireless sensor networks (WSNs) [AKY 02]. In its simplest form, we can describe the network positioning problem as follows.

Consider a two-dimensional (2D) area A with dimensions h and l, with a static sink S, placed in the center of the sensing area A. Sensors should be spread over A and each sensor i is responsible for monitoring a portion of A and periodically reports a series of sensed values to S. Sensors are homogeneous and capable of communicating with S either directly or via multi-hop communication through neighbor sensors. For the simplest case, we can ignore collisions and transmission errors. Each communication has an energy cost, depending on the distance to the next hop, and nodes have a limited amount of energy they can spend through their lifetime. The problem consists of finding the best placements of sensor nodes over the area so that we can maximize the coverage of A and, at the same time, the lifetime of the network.

Simple GA, in the general case, converges to a single solution point. Problems with multiple, often conflicting, objectives require the use of multi-objective GAs [HOR 94]. When evaluating multi-objective optimization problems, the solutions are, normally, a trade-off between the various objectives. In fact, we search for non-dominating solutions or Pareto optimal solutions. A Pareto optimal allocation [YAN 01] is the one where no one could be made better-off without making someone else worse off. In other words, a Pareto allocation is a fair equilibrium point. It is the best allocation we can expect to reach when considering multi-objective functions. It is a point where any change could hurt some of the participants, or objectives in this case. Figure 1.3 presents the set of solutions as defined in [KON 10].

Khanna et al. [KHA 06] present a GA-based allocation algorithm for clustered networks. They are interested in networks that present a hierarchy where the nodes are organized into clusters. The goals of the work are twofold: the first objective is to minimize the power consumption by generating the optimal number of clusters, and cluster members. The second objective is to maximize the sensor coverage. Khanna et al.'s solution consists of two concurrent fitness functions working on distinct problems: one for the node selection and the other for route optimization. Each one of the distinct objectives has its own fitness function and gene representations.

On the nodes selection part of the solution, nodes may be in one of the four states, codified with a 3 bits representation: inactive (000), cluster head (001), inter-cluster router (010) and sensors (100). All nodes in the network have an identification (ID), and the gene is basically the role each node performs in the network. For representing a 25-node network, each gene requires $3 \times 25 = 75$ bits. For example, the representation for the first four nodes, where the first node is a cluster head, the third node is an inter-cluster router and the other two are sensor nodes, would be "001 100 010 100". The genetic operations for node selection are performed over this representation. The fitness function is a weighted function that takes into account four aspects linked to the cluster formation. The four defined partial fitness functions are cluster-head fitness (CHF), node communication fitness (NCF), battery status fitness (BF) and router load fitness (RLF). The total node fitness (TNF) is defined as:

$$TNF = \alpha_1 CHF + \alpha_2 NCF + \alpha_3 BF + \alpha_4 RF \qquad [1.1]$$

where $\alpha_1 + \alpha_2 + \alpha_3 + \alpha_4 = 1$ and α_i depends on the relative significance of the component regarding the final objective.

Nodes periodically transmit status information to the sink, and the sink is responsible for centralizing the information and applying the genetic selection mechanisms. The node selection genetic mechanism runs every time an observed trigger event is reached. Triggers are related to battery alert, deteriorating route fitness alert or a periodic maintenance process. When the optimal fitness is found, the corresponding topology is transmitted to the sensors, which in turn assume the new roles. The genetic operators for the node selection GA are defined as follows:

– *initial population*: the initial chromosomes generated partially randomly and partially based on the previous generation of the network;

– *evaluation*: chromosomes are evaluated using the TNF fitness function ([1.1]);

– *reproduction*: the most suited individuals are arranged first, which provides them a better chance for contributing with offspring to the next generation of the GA. N random elements are selected to a mating pool. From this pool N, new chromosomes are created by rearranging segments from the selected parents. For each offspring, two parents are picked from the pool, and multiple crossover points are randomly selected. The offspring is the mix of these multiple crossover points;

– *mutation*: the new N chromosomes are transferred to the mutation pool where mutation may or may not occur over them. The mutation occurs by changing a bit in the representation string, according to the mutation probability. The mutation probability is adaptive, and is inversely proportional to the average fitness of the population. The bigger the fitness is, the smaller is the mutation rate;

– *selection*: chromosomes are evaluated using the TNF fitness function. The $2N$ existing chromosomes (parents and offspring) are ranked by their fitness and only the best N are chosen to the next generation.

The second objective of Khanna *et al.*'s method is to create well-balanced routes, and the route selection mechanism fulfills this objective. The genes representation for the route selection function also consists of a string of bits, but now, for each node the available connections are represented by zeros and ones. Zeros represent the links that are not used and ones represent the outgoing links of the node. For example, if node 1 can communicate with other four nodes, but the GA decided that it will be able to communicate just with the second one of them, the representation of the routes for the node will be "0100". The fitness function for the route selection takes into account the NCF, BF and the bit rate that is handled by the intercluster router.

The route selection GA presents genetic operators that are similar to the ones on node selection GA. However, it constantly evaluates the network topology to ensure that the route loads are within a reasonable threshold during the usage of the network. Both GA algorithms run in parallel and the status is changed when both agree on a configuration that presents a reasonable convergence point.

Konstantinidis *et al.* [KON 10] proposed another evolution-based solution for the positioning problem. In this work, the authors considered the area A as a regular lattice where nodes can be placed at specific positions. The network placement is defined as a multi-objective function, where one needs to optimize two parameters, energy and coverage. Unfortunately, these goals, coverage and network lifetime, are conflicting objectives. Highly available networks, in general, present a poor coverage, while maximal coverage networks present poor lifetime. The network lifetime, according to Konstantinidis *et al.*, is defined as the duration from the deployment of the network to the time the first sensor completely depletes its energy supply.

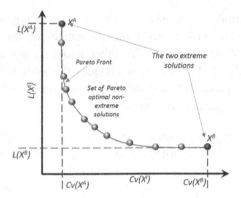

Figure 1.3. *Pareto front solution set for [KON 10]*

As part of their solution, Konstantinidis *et al.* defined two extreme solutions X_A and X_B. X_A provides the maximum lifetime and minimum coverage among all the solutions, while X_B provides the maximum coverage and minimum lifetime. The goal of the proposed method is to present users with a diverse set of Pareto optimal network design choices, giving the trade-off between the extreme solutions X_A and X_B. Pareto optimal solutions close in the objective space, in general, present similarities with each other also in the search space. This characteristic is called proximate optimality principle (POP) [GLO 98]. Konstantinidis *et al.* used this principle as a heuristic on the search for solutions. The multi-objective placement problem is decomposed into *m* scalar subproblems by adding different weights, λ, to each one of the objectives. λ varies from 0 to 1 and is uniformly distributed among the m samples where $\lambda^i = 1-(i/m)$. The first generation of *m* solution sets is generated by randomly picking a series of positions for the sensors over the area. Then, adapted versions of the genetic operators are applied to try to consider the multi-objective problem.

Urrea *et al.*'s [URR 09] main objective is to have a uniform nodes distribution over a given area, but they observe the nodes placement from a distinct angle, as they consider that nodes are mobile. Urrea *et al.* use GA to control the nodes movement targeting to improve the network coverage. The network is divided into a regular hexagonal lattice, where nodes cover a set of cells. The algorithm is based on discrete time steps.

At each time, nodes may move from one cell to any one of the six adjacent cells. The GA controls the movement of the nodes. Nodes exchange genetic information with other nodes in the communication range. The exchanged information includes the node's location, speed and direction. Nodes adapt their own speed and direction according to the instantaneous state of the neighbor nodes.

The method implies that three main objectives need to be observed for achieving the uniform distribution of mobile agents over a given area. The first objective is to maximize the normalized area coverage, defined as the ratio between the total area covered by at least one agent and the overall area. The second objective is to maintain the network connectivity by preventing disconnected nodes. The network is fully connected if all nodes are reachable by any other node through direct or multi-hop communication. The third objective is to provide a balanced and proportional number of neighbors, depending on the network density and agents' positions. An agent located on the middle of the network is expected to have more neighbors, a higher degree, than the nodes at the borders of the observed area. Urrea *et al.* conjecture that these three objectives can be achieved if the node maintains its node degree approximately around an analytically defined mean degree that varies according to the network density, i.e. the area size, number of nodes and the communication range. The analytical mean node degree is used as the fitness function.

Each chromosome represents one node and is made up of three genes. The three genes encode the current location, direction and speed of the mobile node. For each generation, the population is evaluated with respect to the degree of each node. The most suited chromosomes are then chosen for reproduction. After the creation of the next generation, nodes decide their direction and speed based on the knowledge they acquired from the nearby nodes. The most fitting nodes are defined according to their expected analytical degree $\bar{N}$. The node fitness is 1 if and only if it is in the range $\bar{N} - 1$ to $\bar{N} + 1$; any other degree means that the fitness of the node will be smaller than 1. This is important because the method is subdivided into three different cases; case 0 is a baseline for comparison and does not use GA.

In case 1, when two nodes have been selected for reproduction, the mobile node with the lower fitness value adapts the chromosomes to the fitter one, i.e. it adapts to the speed and direction of the most fitting node. In case 2, mobile nodes, selected to reproduction, adjust their speed and direction based on the exchanged genetic material, but the node with better fitness slows down while the node with lower fitness speeds up. If the mobile node reaches the fitness of 1, it stops moving. If in the future the node fitness decreases, e.g. a neighbor node changed position, the node may move again. This approach is the one that reaches the best results from all the evaluated approaches.

In case 2, nodes may be grouped and then they move together. For case 2, nodes are classified as leader, follower or non-follower depending on their current relative positions with respect to their neighboring agents. A node is considered a leader if it has the optimum number of neighbors. It is a follower if it is in the vicinity of a leader, and a non-follower otherwise. A non-follower mobile agent becomes a leader if (1) its fitness value is one, (2) none of its neighbors is a leader and (3) it does not have any follower neighbor. Once a node becomes a leader, it forms a cluster with its followers. The followers adjust their relative positions around the leader so that they are equidistant from it. Once a cluster is formed, its members never move again.

1.4.2. Routing

Routing is one of the most basic tasks in a collaborative computer network. Briefly, routing is the act of finding a path capable of delivering a package between two nodes. Routing is the mechanism that lets us deliver messages from a node source to a node destination. Robust and efficient routing protocols are required in any multi-hop network, and above all for wireless networks. In a fairly interconnected network, given the potential intermediate destinations a packet may traverse before its final destination, network routing may become a complex task [THO 98]. The process becomes even more complex if we consider that not necessarily all parts of the information traverse the same nodes. In general, the data transmitted in the network are often broken into smaller pieces called packets or datagrams. Even if part of the same initial chunk of information and with the same source and destination, depending on the routing strategy, packets may traverse different nodes in the middle. Many algorithms have been proposed for solving the routing problem in its most distinct forms.

In general, routing algorithms associate a weight to each communication link and try to optimize the routing with respect to either one parameter or a set of parameters. According to the most basic definition, routing tries to minimize the number of hops, or links, between a source and a destination, i.e. the shortest path. This problem was initially proposed by Dijkstra [DIJ 59] and is a widely researched subject. Dijkstra's solution is still one of the most efficient methods to solve the shortest path problem. The algorithm consists of organizing the vertices in increasing order of distance from the source and then constructing the shortest path tree edge-by-edge by adding to the tree the available vertex with the smallest path cost each time. The algorithm is simple and elegant; however, its practical implementation raises some issues when the network starts to grow and becomes more dynamic. The tree needs to be constantly recalculated, and this may present a prohibitive cost, and implies that the nodes interconnection information is always available and up-to-date. Taking this into account, Gonen [GON 11] presented a GA-based solution for the shortest path problem.

In Gonen's method, the initial population of the GA is created by randomly generating connection trees from the source to a destination. The fitness function is given by computing the cost of each individual path. The best ranked individuals, the individuals with the lower cost, are chosen for reproduction. The crossover happens by a single crossing point between two randomly selected high ranked nodes. The cross point is also randomly chosen among all the vertices the two parents have in common. If the offspring are fitter than the present high ranked nodes, then they are added to the pool. Otherwise, they are replaced by the fitter solution. The stop condition (when the algorithm stops/ends) for Gonen's work is given by a fixed number of generations, as the objective of the work is to have a reasonable cost path in a limited time, not necessarily the best one.

However, more than just, or exclusively, the number of nodes on the path, many other parameters can be considered. Examples include energy, congestion, delay and jitter. These are all valid metrics that can be considered either alone or in conjunction with each other in the optimization of the routes during the routing process. For example, Baboo and Narasimhan proposed the genetic algorithm-based congestion aware routing protocol (GA-CARP) for mobile *ad hoc* networks [BAB 12]. In their solution, the authors proposed variable length chromosomes that represent the path between a given source and a destination. The initial routes are created by randomly adding connected nodes to the chromosome that represents the path. To avoid loops, for each path nodes can be considered only once. If selected, they are removed from the pool. The process continues until the destination node is reached. The fitness function is given by the links quality, medium access control overhead and the data rate for each chosen link. High-quality routes have a better chance to be chosen for the reproduction process, and thus to have its genetic material preserved into the next generation. The route undergoes mutation by flipping one of the genes of the candidate routes.

Bari *et al.*, on the other hand, concentrate on the problem of creating energy-efficient routes over two-tiered hierarchical sensor networks [BAR 09]. Their approach is capable of increasing the network lifetime for an impressive 200%, when compared to other traditional routing schemes. Moreover, for small networks, where the global optimum can be determined, the method is always capable of finding the optimal solution. Often, in a sensor network, the lifetime of the network relies on the battery power of few critical nodes [AKK 05]. In this type of scenario, the routing protocol can play a major role in increasing the lifetime of the network.

The problem may become even more interesting if we consider that the nodes are not homogeneous, and that special high-powered relay nodes may be used to concentrate and redistribute the information. The target topology in [BAR 06] is a hierarchical two-tired network where higher powered relay nodes act as cluster heads and sensor nodes transmit their data to these cluster heads. These are then responsible for transmitting the message to the sink through multi-hop or direct transmission. Each sensor node belongs to one single cluster, i.e. it can connect to only one cluster head. The complete drain of power over a relay node may have a severe impact on the network, since the nodes belonging to its cluster will be unable to send their data to the sink, i.e. a portion of the target area will be uncovered due to the failure of one single node. Bari *et al.* developed a centralized GA to try to decrease this problem where the sink node, which has no power constraints, is responsible for running the algorithm and defining the routes over the network. Network nodes are considered static or with very low mobility. The routing problem can be influenced by the way nodes are distributed in the area. For example, in a sensor network, nodes can be placed either at specific positions or spread over the terrain. In the first case, the routing decision can be done even before the deployment of the network, and even possibly influence the decision of where to place the nodes, i.e. the topology may adapt to the routing. In the second case, the nodes' position is not predetermined, and the actual positions can be found by a different method, e.g. a global positioning system (GPS) and the clusters assigned by other distinct clustering methods. In [BAR 09], the network lifetime criterion is N-of-N, which means that the network lifespan is given by the number of rounds until the first relay node fails due to the lack of power [PAN 03]. Chromosomes are represented as string of nodes IDs. The length of each chromosome is always equal to the number of relay nodes and the position on the chromosome; the ith position represents the next hop of node i.

The initial population is formed by valid routing schemes where the routes are created as connected trees from the sink node. The projected total number of rounds until the first relay node runs out of battery is used as the fitness function. The fitness for a given individual is computed as $L_{net} = E_{initial}/ E_{max}$, where L_{net} is the network lifetime in terms of rounds, $E_{initial}$ is the initial energy of a relay node and E_{max} is the maximum energy dissipated by a relay node on the round of data gathering. The proposed scheme is also capable of dynamically adapting to the status of the network. The fitness function takes into account the nodes' residual energy and the expected cost in energy for a given route scheme, defined on each of the individual's chromosome.

It is common that over the network lifetime, different relay nodes present different residual energy levels. Even if the routing algorithm is correct and balances the charges well, no routing scheme is resilient to unforeseen events, such as node failures for example. In [BAR 09], new routing schemes are started periodically, but they are also triggered by network failures. The GA can even continuously run on the server, observing the residual energy reported periodically by the relay nodes. In this case, the new routing scheme would be released whenever a reasonable benefit is found.

Selection is carried out by using the Roulette-Wheel selection method [GOL 89] in which the most fitting nodes have a higher chance of being selected for reproduction. The crossover is done by randomly selecting k-points crossover [GOL 89], where k varies from 1 to 3. After the crossover process, two new offspring, which are a mix of the k-points, are generated. The mutation operator is distinct from the regular random operator. In [BAR 09], the gene, i.e. node, chosen to suffer a mutation is the one that uses the most energy regarding its reception/transmission profile. The objective is to decrease the overall need of energy from the scheme, thus increasing the network lifetime. This can be done in two ways. First, by replacing an edge from the critical node and then moving it to another closer node. This would reduce the energy requirements for the node since the next node would be closer. The second way is by diverting one incoming flow to another node, alleviating in this way the traffic over the critical node. Figure 1.4 presents these two alternatives.

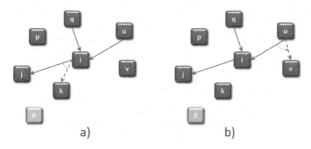

a)   b)

Figure 1.4. *Redistribution of load on critical node as mutations [BAR 09]*

Yen *et al.* [YEN 08] also proposed an energy-efficient routing scheme based on GA, but in their case they were interested in the multicast routing problem. The authors developed a source tree-based routing algorithm building the shortest path multicast tree to minimize the delay time and increase the network lifetime. The chromosomes encoding method consists of two lists of integers. The first chromosome, S, encodes the position of the node on the tree, and consists of a string of nodes IDs. The second chromosome, T, codifies the gene topology and encodes the parent of the respective node on chromosome S. With S and T, the algorithm is capable of recreating the multicast tree. Figure 1.5 presents the codification scheme for a nine-node network, as described in [YEN 08]. The initial population consists of feasible random tree topologies. The fitness function is based on penalty relations regarding a number of parameters such as delay, remaining energy at node, battery capacity, transmission power, node degree and cost to repair a broken route. The tree node with the smaller residual energy defines the multicast tree maximum lifetime. The crossover is done by selecting the two chromosomes with greater fitness values on the chromosomes pool and randomly selecting the point and length of the chromosome to be exchanged between parents. After creation, the two new offspring added to the chromosomes pool.

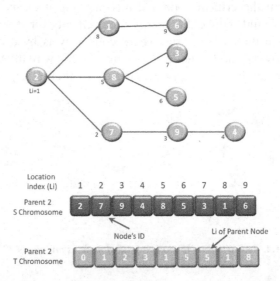

Figure 1.5. *Extended sequence and topology encoding for MTree [YEN 08]*

The mutation operation in [YEN 08] is also biased by the energy constraints. The residual battery energy acts as a mutation factor. Nodes with higher energy, normally the leaf nodes, replace internal tree nodes with lower energy. This makes a well-balanced multicast tree possible, in terms of remaining battery and perspective of use of such energy.

1.4.3. *Other works*

Evolutionary methods are used in a number of papers and problems concerned with computer networks. Among the many available works, we can find the *entropy-based genetic algorithm* to support QoS multicast routing protocol [CHE 06], a genetic approach to joint routing and link scheduling for wireless mesh networks. The hierarchical QoS multicast routing uses a method to compute, in a distributed way, a Semi-Optimal Multicast Tree in mobile *ad hoc* networks, based on GA [TAK 07]. In [KIM 07], GA is used to find efficient network coding solutions. Sharples and Wakeman [SHA 00] developed an automatic protocol synthesis method based on GA. Cache routing strategies are studied in light of GA in [BRA 07]. Alouf *et al.* presented a GA-based method for optimizing forwarding policies of an end-to-end epidemic message relaying protocol [ALO 07]. In this work, the genotype consists of the forwarding probability and the number of messages that should be rebroadcast. The forwarding decisions are based on the values of the chromosomes.

The works discussed here represent by no means an exhaustive collection of all the existent applications of evolutionary techniques in the networks field. The objective of this chapter is to present some examples and discuss the way the evolutionary operators can be used to solve real complex network problems.

1.5. Further reading

Readers interested in learning more about artificial evolutionary techniques can refer to the following works.

EIBEN A.E., SMITH J.E., *Introduction to Evolutionary Computing*, Natural Computing Series, Springer, October 2008.

POLI R., LANGDON W.B., FREITAG MCPHEE N., *A Field Guide to Genetic Programming*, Lulu Enterprises, UK Ltd, March 2008.

ALBA E., TROYA J.M., "A survey of parallel distributed geneticalgorithms", *Complex*, vol. 4, no. 4, pp. 31–52, March 1999

MELANIE M., *An Introduction to Genetic Algorithms*, MIT Press, Cambridge, MA, 1998.

CANTU-PAZ E., "A survey of parallel genetic algorithms", *Calculateurs Parallèles, Réseaux et Systèmes Répartis*, vol. 10, no. 2, pp. 141–171, 1998.

GOLDBERG D.E., *Genetic Algorithms in Search, Optimization, and Machine Learning*, 1st ed., Addison-Wesley Professional, 11 January 1989.

For works on the application of evolutionary methods over computer networks problems readers can consult the following works.

DORRONSORO B., RUIZ P., DANOY G. *et al.*, *Evolutionary Algorithms for Mobile Ad Hoc Networks*, Nature-Inspired, Computing Series, Wiley, 2014.

AGGARWAL C., SUBBIAN K., "Evolutionary network analysis: a survey", *ACM Computing Surveys*, vol. 47, no. 1, Article 10, May 2014.

MAJEED P.G., KUMAR S., "Genetic algorithms in intrusion detection systems: a survey", *International Journal of Innovation and Applied Studies*, vol. 5, no. 3, pp. 233–240, March 2014.

LIO P., VERMA D. (eds), *Biologically Inspired Networking and Sensing: Algorithms and Architectures*, IGI Global, 2011.

REZAZAD H., "Computer network optimization", *Wiley Interdisciplinary Reviews: Computational Statistics*, vol. 3, no. 1, pp. 34–46, January/February 2011.

VENKETESH P., VENKATESAN R., "A survey on applications of neural networks and evolutionary techniques in web caching", *IETE Technical Review*, vol. 26, no. 3, pp. 171–180, 2009

SMITH A.E., DENGIZ B., "Evolutionary methods for design of reliable networks", *Telecommunications Optimization: Heuristic and Adaptive Techniques*, Wiley, pp. 17–34, 2001.

1.6. Bibliography

[AKK 05] AKKAYA K., YOUNIS M., "A survey on routing protocols for wireless sensor networks", *IEEE Transactions on Mobile Computing*, vol. 3, no. 3, pp. 325–349, 2005.

[AKY 02] AKYILDIZ I.F., WEILIAN S., SANKARASUBRAMANIAM Y. *et al.*, "A survey on sensor networks", *Elsevier Computer Networks*, vol. 38, no. 4, pp. 393–422, 15 March 2002.

[ALO 07] ALOUF S., CARRERAS I., MIORANDI D. *et al.*, "Embedding evolution in epidemic-style forwarding", *Proceedings of the 4th IEEE International Conference on Mobile Ad-hoc and Sensor Systems (MASS '07)*, Pisa, Italy, October 2007.

[BAB 12] BABOO S.S., NARASIMHAN B., "Genetic algorithm based congestion aware routing protocol (GA-CARP) for mobile ad hoc networks", *2nd International Conference on Computer, Communication, Control and Information Technology (C3IT-2012)*, *Procedia Technology*, vol. 4, pp. 177–181, 2012.

[BAD 09] BADIA L., BOTTA A., LENZINI L., "A genetic approach to joint routing and link scheduling for wireless mesh networks", *Elsevier Ad Hoc Networks*, vol. 7, pp. 654–664, 2009.

[BAR 06] BARI A., JAEKEL A., BANDYOPADHYAY S., "Optimal load balanced clustering in two-tiered sensor networks", *Proceedings of the 3rd IEEE/CreateNet International Workshop on Broadband Advanced Sensor Networks (BASENETS)*, CA, October 2006.

[BAR 09] BARI A., WAZED S., JAEKEL A. *et al.*, "A genetic algorithm based approach for energy efficient routing in two-tiered sensor networks", *Elsevier Ad Hoc Networks*, vol. 7, pp. 665–676, 2009.

[BRA 07] BRANKE J., FUNES P., THIELE F., "Evolutionary design of en-route caching strategies", *Elsevier Applied Soft Computing*, vol. 7, no. 3, pp. 890–898, 2007.

[CHE 06] CHEN H., SUN B., ZENG Y., "QoS multicast routing algorithm in MANET: an entropy-based GA", *Proceedings of the International Conference on Intelligent Computing*, Kunming, China, pp. 1279–1289, August 2006.

[CLE 99] CLEMENTI A.E.F., PENNA P., SILVESTRI R., "Hardness results for the power range assignment problem in packet radio networks", *Proceedings of the International Workshop on Approximation Algorithms for Combinatorial Optimization Problems*, Springer-Verlag, pp. 197–208, 1999.

[COB 90] COBB H.G., An investigation into the use of hypermutation as an adaptive operator in genetic algorithms having continuous, time-dependent nonstationary environments, Technical Report AIC-90-001, Naval Research Laboratory, Washington D.C., 1990.

[DAR 59] DARWIN C., *On the Origin of Species, or the Preservation of Favoured Races in the Struggle for Life*, John Murray, London, 1859.

[DIJ 59] DIJKSTRA E.W., "A note on two problems in connexion with graphs", *Numerische Mathematik*, vol. 1, pp. 269–271, 1959.

[DOO 00] DOOLITTLE W.F., "Uprooting the tree of life", *Scientific American*, vol. 282, no. 2, pp. 90–95, February 2000.

[EIB 96] EIBERG H., MOHR J., "Assignment of genes coding for brown eye colour (BEY2) and brown hair colour (HCL3) on chromosome 15q", *European Journal of Human Genetics*, vol. 4, no. 4, pp. 237–241, 1996.

[GLO 98] GLOVER F., LAGUNA M., *Tabu Search*, Kluwer, Norwell, MA, 1998.

[GOL 89] GOLDBERG D.E., *Genetic Algorithms in Search, Optimization, and Machine Learning*, Addison Wesley, Reading, MA, 1989.

[GON 11] GONEN B., "Genetic algorithm finding the shortest path in networks", *Proceedings of the 2011 International Conference on Genetic and Evolutionary Methods*, Las Vegas, NV, July 2011.

[HOL 92] HOLLAND J., *Adaptation in Natural and Artificial Systems*, MIT Press, Cambridge, MA, 1992.

[HOR 94] HORN J., NAFPLIOTIS J.N., GOLDBERG D., "A niched Pareto genetic algorithm for multiobjective optimization: evolutionary computation", *Proceedings of the 1st IEEE Conference on Computational Intelligence*, pp. 82–87, 1994.

[KHA 06] KHANNA R., LIU H., CHEN H.H., "Self-organization of sensor networks using genetic algorithms", *IEEE International Conference on Communications (ICC '06)*, Istanbul, Turkey, 11–15 June 2006.

[KIM 07] KIM M., MEDARD M., AGGARWAL V. *et al.*, "Evolutionary approaches to minimizing network coding resources", *Proceedings of the IEEE Infocom 2007*, Anchorage, AL, pp. 1991–1999, 2007.

[KON 10] KONSTANTINIDIS A., YANG K., ZHANG Q. *et al.*, "A multi-objective evolutionary algorithm for the deployment and power assignment problem in wireless sensor networks", *Elsevier Computer Networks*, vol. 54, pp. 960–976, 2010.

[MEN 66] MENDEL G., "Versuche über Pflanzen-Hybriden", *Verh. Naturforsch, Ver. Brünn*, vol. 4, pp. 3–47, 1866 (in English, *J. R. Hortic. Soc.*, vol. 26, pp. 1–32, 1901).

[O'CO 08] O'CONNOR C., "Meiosis, genetic recombination, and sexual reproduction", *Nature Education*, vol. 1, no. 1, p. 188, 2008.

[PAN 03] PAN J., HOU Y.T., CAI L. *et al.*, "Topology control for wireless sensor networks", *Proceedings of the 9th Annual International Conference on Mobile Computing and Networking*, pp. 286–299, 2003.

[PEA 06] PEARSON H., "Genetics: what is a gene?", *Nature*, vol. 441, pp. 398–401, 2006.

[RIE 04] RIESER C.J., RONDEAU T.W., BOSTIAN C.W. *et al.*, "Cognitive radio test bed: further details and testing of a distributed genetic algorithm based cognitive engine for programmable radios", *IEEE Military Communications Conference*, 2004.

[SAN 04] SANJUAN R., MOYA A., ELENA S.F., "The distribution of fitness effects caused by single-nucleotide substitutions in an RNA virus", *Proceedings of the National Academy of Sciences U.S.A.*, vol. 101, no. 22, pp. 8396–8401, 2004.

[SAW 07] SAWYER S.A., PARSCH J., ZHANG Z. *et al.*, "Prevalence of positive selection among nearly neutral amino acid replacements in Drosophila", *Proceedings of the National Academy of Sciences U.S.A.*, vol. 104, no. 16, pp. 6504–6510, 2007.

[SHA 00] SHARPLES N., WAKEMAN I., "Protocol construction using genetic search techniques", *Real-World Applications of Evolutionary Computing*, Lecture Notes in Computer Science, vol. 1803, pp. 235–246, Springer, 2000.

[TAK 07] TAKASHIMA E., MURATA Y., SHIBATA N. *et al.*, "A method for distributed computation of semi-optimal multicast tree in MANET", *Proceedings of the 8th IEEE Wireless Communications and Networking Conference*, Kowloon, China, March 2007.

[THO 98] THOMAS T.M., *"OSPF Network Design Solutions"*, Cisco Press, 1998.

[TUR 50] TURING A., "Computing machinery and intelligence", *Mind LIX*, vol. 59, no. 236, pp. 433–460, October 1950.

[URR 09] URREA E., SAHIN C.S., HÖKELEK I. *et al.*, "Bio-inspired topology control for knowledge sharing mobile agents", *Elsevier Ad Hoc Networks*, vol. 7, pp. 677–689, 2009.

[WHI 98] WHITLEY D., RANA S., HECKENDORN R.B., "The island model genetic algorithm: on separability, population size and convergence", *Journal of Computing and Information Technology*, vol. 7, pp. 33–47, 1998.

[YAN 01] YANG X., *Economics: New Classical Versus Neoclassical Frameworks*, Wiley-Blackwell, 2001.

[YEN 08] YEN Y.S., CHAN Y.K., CHAO H.C. *et al.*, "A genetic algorithm for energy-efficient based multicast routing on MANETs", *Elsevier Computer Communications*, vol. 31, pp. 2632–2641, 2008.

[ZUC 65] ZUCKERKANDL E., PAULING L., "Evolutionary convergence and divergence in proteins", in BRYSON V., VOGEL H.G. (eds), *Evolving Genes and Proteins*, Academic Press, New York, 1965.

2

Chemical Computing

Word cloud representing the full text of this chapter and the words frequencies. Created with Wordle.net

Chemistry is largely about chemical changes and chemical reactions expressing a chemical change [BAL 11]. A chemical reaction consists of the transformation of one set of chemical substances, called reactants, into another set of substances, called the products. These chemical changes involve breaking bonds in the reactants, rearranging the atoms and forming new bonds in the products. Therefore, not only must a collision occur between reactant particles, but it must also provide sufficient energy to break reactant bonds in order to form new products [POU 10]. The product, or products, of a chemical reaction usually has distinct properties than the reactants. A good example is the volcano-like science project, where we mix baking soda (sodium bicarbonate) and vinegar (acetic acid). The reaction produces carbon dioxide gas; if we add dish soap to the mixture, the release of dioxide gas will create a large amount of bubbles.

It is important to highlight here the distinction between chemical and physical changes. For example, the change in temperature is a physical change. When the snow melts and turns into water, this is a physical change and the chemical compound (H_2O) remains the same. On the other hand, when a metal gets rusty, this is a chemical reaction. The iron (Fe) in the metal combines with the oxygen (O_2) in the air to form another product called iron oxide (Fe_2O_3). A chemical change must occur into a reaction, and they are usually harder to reverse than the physical changes.

Changes can occur among atoms, ions, compounds or molecules of complex or single elements. We can represent chemical changes more succinctly as equations in the form "hydrogen + oxygen → water", or using the symbols that represent each one of the involved elements and their quantities, "$H_2 + O \rightarrow H_2O$". The same amount of elements in the first part needs to be represented in the second part; no new atoms are created, and no atoms are destroyed. This means that the total mass of the reactants must be equal to the mass of the products, as stated by the law of conservation of mass [POU 10].

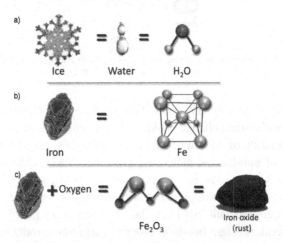

Figure 2.1. a) *Chemical structure of water (H_2O); the chemical structure does not change independently of its state, solid or liquid, i.e. it is a physical reaction and not a chemical one. b) The structure of iron and c) the combination of iron and oxygen modifies the structure, creating another compound, i.e. it is a chemical reaction*

Reactions often occur in a series of one-step smaller reactions; each of these individuals is called an elementary reaction. An elementary reaction is a single-step reaction with a single transition state and no intermediates [NGU 14].

Each of these steps uses a series of reactants, or intermediates, if the elementary reaction uses by-products of previous elementary reactions. However, it is important to note that reactions may not always use all the available reactants. In fact, reactions may have three possible outcomes: first, they may go to the full completion, i.e. in the end the reaction vessel contains products, and only products. Second, the reaction may not start at all; so at the end the reaction vessel contains all reactants, and only reactants. Third, some reactions may start but not go to completion, that is the reaction might start but not go completely to products. In this case, the reaction vessel would contain reactants and products in different quantities. Moreover, reactions may be irreversible or reversible. When a reaction goes only in one sense, i.e. the reactants form products and these products are stable, there is no going back; we call this an irreversible reaction. In general, this kind of reaction keeps going until all the reactants have been used up and there is not any more left. It is the first reaction case. In a reversible reaction, the reaction can go forward and backward, i.e. products may react to form reactants. This kind of reaction may lead to the third possible reaction scenario in which we have a mix of products and reactants in the reaction vessel.

A reversible reaction is said to be in equilibrium, or in dynamic equilibrium, if the forward and reverse paths happen at the same rate. At this point, the concentrations of reactants and products do not change. The reaction appears to have stopped before one of the reactants has run out. However, the reactions are still happening in both senses; the only thing is that they are happening at the same, or nearly the same, rate. The concentrations of reactants and products are not necessarily the same between them. A reaction in equilibrium means only that the forward and reverse reactions happen at the same rate and that the individual concentration of reactants and products does not change significantly.

2.1. Artificial chemistry

Artificial chemistry (AC) is a technique inspired by real chemistry. It has emerged as a subfield of the artificial life research field as an extreme bottom-up approach to artificial life. Back in the 1950s, Turing already called attention to the influence of chemical reactions and mathematical abstractions on modeling life-related process [TUR 52]. The fundamental idea behind AC is that if we would want to build something alive, the starting point would be to combine non-living entities. Some researchers in the artificial intelligence (AI) field believe that the explanation to life passes through the concept of emergency [DIT 01]. Emergency is the process in which large complex entities, or patterns, may emerge from the interactions among smaller and simpler elements. These building blocks do not necessarily present the same characteristics found in the more complex structures, and this is exactly what happens in chemistry. For example, two gases, oxygen and hydrogen, at normal conditions of temperature and pressure, combine to form a liquid, water. In this case, what changes is only the way the atoms are linked. The properties of the product changed not because we changed the reactants, but because they are simply organized in a different manner. We could speculate that this kind of emergence could happen in other much more complex systems. It may even happen in living organisms, which are alive not because of the atoms that compose them, but because the way these atoms are organized [DIT 01].

AC is based on the principles of traditional chemistry. As such, it focuses on the fundamental process of combining elements, and, in particular, creating new ones. AC studies forms of organization, self-maintenance, self-construction and the conditions in which this happens. Artificial chemical computing is the branch of AC that deals with the use of the molecules to try to perform calculations.

We can say that AC is about combining elements that change under certain circumstances and, in particular, about systems that can build new components. Thus, AC deals with the basic conditions required to build an environment where forms of organization, self-maintenance and self-construction can arise [DIT 01]. In an unconventional way, it presents all the required elements for a universal computation environment. Chemical reactions are responsible for carrying out the computations where the basic computational elements are the molecules present in the system, which can be consumed, or transformed, to produce new computational elements. Important elements in the computation process end up with higher concentration, and less important elements end up with lower concentration or no concentration at all. Indeed, in AC, the concentration of the elements determines the kind of computation needed to be performed [BAN 04].

Usually, AC is defined as a triple (S, R, A), where S is the set of all possible molecules, where the number of elements may be infinite. R is the set of collision rules among the molecules in S. R is normally represented by a chemical reaction (e.g. A+B $\rightarrow$ C+D), where A is the algorithm that describes how to apply R over a set of elements from S [DIT 01]. Alternatively, we could represent AC as the tuple (S, I), where S is the set of molecules and I are the interaction rules over S. In some cases, this simplified representation may be useful.

The way the rules are applied over S can be divided fundamentally into two ways: either each molecule collision is treated explicitly or the reactions are represented by the frequency and the concentration of molecules on the reaction vessel. Stochastic molecular collisions, when we treat each collision, randomly draw a sample of molecules from S and verify if any of the rules in R applies. If so, the reaction products substitute the reactants. If more than one rule applies, a decision is taken in terms of precedence. The products or reactants to which no rule applies return to the vessel. This process continues until the termination condition is reached. The advantage of this process is that it is similar to what happens in the real world and is easily parallelizable. However, if we take into account the possible number of molecules and reactions, the cost in terms of processing may be very high. This may become even worse if the concentration of one set of molecules is orders of magnitude smaller than that of others. In this case, we may keep a log time processing molecules that cannot react with each other.

However, discrete difference equations do not treat each molecule separately. It handles the concentrations of the elements, taking into account the dynamics of the chemical reactions. The concentrations are calculated through differential equations in which the functions represent physical quantities and their derivatives represent the rates of change. The equations define the relationship between the quantities and the changing rate. The metadynamics method [BAG 92] admits that the concentration rates can change during the reaction process. Moreover, the number of differential equations, and their concentration thresholds, may also change over time. The concentration threshold defines whether the equation is applicable or not. Below the threshold value, the equation is not applicable anymore.

2.2. Applications on networks

AC approaches require a paradigm shift in the way we express computation. The way programs are developed and executed differs greatly from methods that are more traditional. In regular programming languages, instructions are typically executed sequentially; however, this is not true in AC. In traditional, mono-processor machines, the code is executed instruction after instruction in a precise order previously defined by the programmer. In AC, the instructions are executed whenever possible and in accordance with their concentration on the reaction vessel. Thus, AC can be understood as a massive parallel machine with no prior scheduling of instructions. At each instant in time, a set of instructions is obtained from the pool and executed independently.

2.2.1. Data dissemination

Miorandi et al. [MIO 11] investigate the use of AC on the data-centric message distribution problem over delay-tolerant networks (DTNs). DTNs are the kind of network where we have intermittent connection between nodes. In general, messages are retransmitted between nodes in an opportunistic and forward way. This means that the nodes store the messages they received and whenever a node encounters another node, it can decide to forward the message or not.

In the general case, DTNs are the *best effort* networks, i.e. there is no strong guarantee that a given message will reach the destination.

In their work, Miorandi *et al.* argue that the best way to disseminate messages in a DTN is with data-centric network techniques. In data-centric dissemination systems, the content is the most important. The interest of the users on that specific piece of information is what drives the messages' dissemination and, as such, their availability on the network. Routing paths and nodes' addresses, even what was the original source of the packages, are of little importance. In [MIO 11], each message is treated as a molecule, which travels among the nodes. Each message has a semantic description of the set of reactions it can participate in. Nodes are considered chemical reactors and, from time to time, they broadcast to neighbor nodes requests for molecules of interest. If a node that hosts one of these molecules receives the message of interest, this triggers a reaction to deliver a copy of the requested content. Each molecule has a content tag, which is a semantic description of the message content and defines the reactions in which the molecule may be involved.

The data management process is implemented as a "dilution flux", which tries to control the number of messages/molecules available over the network of reactors (nodes). The data are exchanged in a collaborative way, where nodes relay the messages until someone has taken interest in it. The diffusion of the data is based on the interest it represents to the nodes. The higher the interest, the more spread a given content will be.

Miorandi *et al.* used the Fraglets language [TSC 03] and simulations to prove the feasibility of their technique. In their paper, the authors argued that the Fraglets language is particularly interesting to implement AC over computer networks because, first, it was exactly developed to express network protocols and, second, the way it expresses code and data makes it easy to implement code mobility. A fraglet is expressed as a function of $n[a\ b\ tail]x$, where n represents the node identifier where the fraglet is present, a, b and tail are the symbols and x denotes the multiplicity of the fraglet. These components represent the way the fraglet will execute over the nodes. We can say that they would be the lines of code, in a rough analogy to standard programs. The advantage of fraglets is that they can be combined to create self-evolving and self-healing programs. In our case, fraglets become the molecules of the AC method. As it happens with the molecules in traditional chemistry, fraglets apply computation over other fraglets to create new fraglets. Simple reactions lead to a series of fraglets' transformation, which includes their transfer to another node. Complex reactions combine two fraglets to create a new fraglet. With these reactions, the system evolves and the elements may change. If, at a given time, more than one reaction is possible, the system chooses randomly which reaction should be executed. The involved fraglets are then removed from the reaction vessel, and processed, and the results return to the vessel. Fraglet reactions are highly parallelizable; the only dependency is the removal of the reactants from the vessel each time to avoid changing the concentration. Readers are advised to refer to [TSC 03] for a full fraglet evolution example.

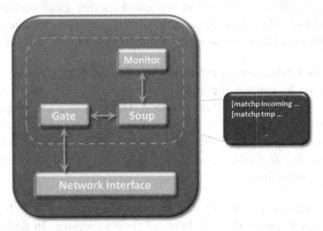

Figure 2.2. *Internal architecture for the simulated node in [TSC 03]*

2.2.2 *Routing*

For Kephart and Chess, computational systems should manage themselves autonomously and new components should be integrated to the already established system as simply as "a new cell establishes itself in the human body" [KEP 03]. In the beginning of the 2000s,

IBM perceived that computer systems were starting to be too complex to be efficiently managed, and from this arose the *Autonomic Computing* concept, as part of a movement inside IBM to create auto-managed systems. From the autonomic computing point of view, systems should be capable of self-configuration, self-healing, self-optimization and self-protection, which is commonly called self-X.

The ideas of Kephart and Chess inspired the creation of a self-healing routing algorithm, inspired by AC [MEY 08]. Given its source of inspiration, the proposed protocol is highly resilient and can even recover from random removals of its own instructions. The intention of the authors is to create a software that will be able to detect and repair erroneous code independently if such a code is the result of programming mistakes or execution using an unreliable transmission medium.

Unlike traditional routing algorithms, where explicit metrics control the data flows, in [MEY 08] with dictates, the path of a message through the network is their concentration on the routing tables. Rules compete with each other to deliver the packets. A reinforcement mechanism rewards successful forwarding rules, which increase their concentration and as such their importance.

The routing algorithm is a cooperative self-replicating system, where its elements constantly compete for the limited resources available. Parts of the system constantly regenerate themselves. It is from this mechanism that the self-healing property of the protocol arises. Damaged parts, which are not capable of self-replication anymore, disappear from the system through the dilution process flow, and are replaced by healthy duplicated elements.

The protocol consists of two interactive subsystems: route dissemination and regulation. Both use code self-replication to withstand code deletion attacks. As in [MIO 11], here the molecules are also expressed with fraglets. Each node is an independent artificial chemical reactor, described by its own triple (Si, Ri, Ai), where Si is the set of molecules in node i, Ri is the set of collisions on node i and Ai is the application of Ri over Si at node i. The continuous monitoring and self-healing of the system requires the program to be able to modify its own code at run time, which can hardly be done with traditional static programs. Here is where it becomes an apparent advantage of AC and fraglets over traditional methods. The system is based on the structure of eukaryotic cells, with a nucleus vessel that handles the routing table entries and another vessel, the main vessel, that is responsible for the packet forwarding activities.

The dilution process is responsible for continuously eliminating molecules to make room for new ones. In fact, we can say that the system continuously regenerates as the operational blocks replicate their own code while being executed. Whenever the required input molecules are available (requests for a given piece of information, service or route), the reaction rules are triggered, and the desired computation is performed. The reaction produces both the molecule that represents the reaction and a replication reward, meaning the duplication of the used rule. In practice, this means that when selected randomly, reactions with higher concentration have more chances of being applied than others. If the required molecules are available, the inputs and the number of rules to treat them will grow exponentially since the rule is duplicated each time it is applied. With the decrease in the number of reactants required to this specific rule, its concentration tends to decrease for two reasons. The first reason is the dilution process and second reason is the augmentation in concentration of the rules used to treat a new set of input molecules. In practice, the dilution flow works by maintaining the number of molecules in the reaction vessel smaller than a given maximum. At the end of each reaction interaction, the system verifies the number of molecules that exceed the capacity of the vessel and randomly removes molecules from the vessel until the number of exceeding moles is zero.

Instead of treating each and every rule, the Gillespie's algorithm [GIL 77] is used to stochastically simulate the reactions according to their concentration rates. For each interaction, the algorithm calculates the reactions that will take place based on a probabilistic selection. The routing rules, i.e. molecules, are then applied and the packets are forwarded to the next node capable of threatening the request, or delivering the information. The forwarding rules are chosen matching the packet's destination and are based on the corresponding rules concentrations. When a packet reaches the destination, an acknowledgement packet is sent back along the reverse path. This packet reacts with all corresponding rules used in the forwarding of the initial packet and triggers the replication of the rule. This increases the concentration of successful rules. This replication mechanism is quite efficient, but as the vessel has a size limit and old rules need to disappear to make space for new ones, this could overwrite important but little used rules. To overcome this, and to ensure the diversity of the available rules, the nucleus vessel periodically recreates copies of the available routes, even ones used infrequently. This yields the robustness of the algorithm and the availability of rules if the concentration of the inputs changes. For example, imagine that inside the rules we have two paths to the same destination. One path is quite advantageous and the other path is much less interesting. The concentration of the first path will be much higher than the second path; however, the nucleus will ensure that the second path will also have some representatives on the reaction vessel. If for some reason the second path suddenly becomes more attractive, its concentration will increase leading to an equilibrium across the available options.

2.3. Further reading

Readers interested in learning more about artificial chemistry can refer to the following titles.

DITTRICH P., ZIEGLER J., BANZHAF W., "Artificial chemistries – a review", *Artificial Life*, vol. 7, no. 3, pp. 225–275, 2001.

HUTTON T.J., "Evolvable self-replicating molecules in an artificial chemistry", Artificial Life, vol. 8, no. 4, pp. 341–356, September 2002.

ERCIM, Unconventional programming paradigms, workshop report, ERCIM, Mont Saint-Michel, France, 15–17 September 2004.

BANÂTREA J.P., FRADETB P., RADENAC Y., "A generalized higher-order chemical computation model", *Proceedings of the 1st International Workshop on Developments in Computational Models (DCM 2005), Electronic Notes in Theoretical Computer Science*, vol. 135, no. 3, pp. 3–13, 3 March 2006.

DITTRICH P., FENIZIO P.S., "Chemical organization theory: towards a theory of constructive dynamical systems", *arXiv:q-bio/0501016 [q-bio.MN]*, February 5 2008.

For works on the application of AC methods over computer networks problems the reader can consult the following publications.

MIORANDI D., CARRERAS I., DE PELLEGRINI F. *et al.*, "Chemical relaying protocols", *Bio-Inspired Computing and Communication Networks*, Taylor & Francis Group, CRC Press, 2011.

TSCHUDIN C., "Fraglets – a metabolistic execution model for communication protocols", *Proceedings of the 2nd Annual Symposium on Autonomous Intelligent Networks and Systems (AINS)*, Menlo Park, CA, 2003.

YAMAMOTO L., TSCHUDIN C., "Experiments on the automatic evolution of protocols using genetic programming", *Proceedings of the 2nd IFIP Workshop on Autonomic Communication (WAC)*, Athens, Greece, 2005.

2.4. Bibliography

[BAG 92] BAGLEY R.J., FARMER J.D., "Spontaneous emergence of a metabolism", *Artificial Life II*, Addison-Wesley, Redwood City, CA, pp. 93–140, 1992.

[BAL 11] BALL D.W., Introductory Chemistry, Saylor Foundation, available at http://www.saylor.org/books/, 2011.

[BAN 04] BANZHAF W., LASARCZYK C., "Genetic programming of an algorithmic chemistry", O'REILLY U.-M. *et al.* (eds.), *Genetic Programming Theory and Practice II*, Kluwer/Springer, vol. 8, pp. 175–190, 2004.

[DIT 01] DITTRICH P., ZIEGLER J., BANZHAF W., "Artificial chemistries – a review", *Artificial Life*, vol. 7, no. 3, pp. 225–275, 2001.

[GIL 77] GILLESPIE D.T., "Exact stochastic simulation of coupled chemical reactions", *The Journal of Physical Chemistry*, vol. 81, no. 25, pp. 2340–2361, 1977.

[KEP 03] KEPHART J.O., CHESS D.M., "The vision of autonomic computing", *Computer*, vol. 36, no. 1, pp. 41–50, 2003.

[MEY 08] MEYER T., YAMAMOTO L., TSCHUDIN C., "A self-healing multipath routing protocol", *Proceedings of the 3rd International Conference on Bio-Inspired Models of Network, Information and Computing Systems (BIONETICS '08)*, Institute for Computer Sciences, Social-Informatics and Telecommunications Engineering (ICST), Hyogo, Japan, 25–28 November 2008.

[MIO 11] MIORANDI D., PELLEGRINI I.C.F., CHLAMTAC I. *et al.*, "Chemical Relaying Protocols" in XIAO Y., HU F. (eds), *Bio-inspired Computing and Communication Networks*, Auerbach Publications, Taylor & Francis Group, CRC Press, 2011.

[NGU 14] NGUYEN T., NGO M., Elementary reactions, available at http://chemwiki.ucdavis.edu/Physical_Chemistry/Kinetics/Rate_Laws/Reacti on_Mechanisms/Elementary_Reactions, 2014.

[POU 10] POULSEN T., Introduction to chemistry, Open Education Group, available at http://openedgroup.org/publications, 2010.

[TSC 03] TSCHUDIN C., "Fraglets – a metabolistic execution model for communication protocols", *Proceedings of the 2nd Annual Symposium on Autonomous Intelligent Networks and Systems (AINS)*, Menlo Park, CA, July 2003.

[TUR 52] TURING A.M., "The chemical basis of morphogenesis", *Philosophical Transactions of the Royal Society of London, Series B: Biological Sciences*, vol. 237, no. 641, pp. 37–72, 14 August 1952.

3

Nervous System

*Word cloud representing the full text of this chapter and the
words frequencies. Created with Wordle.net*

The Nervous System (NS) is responsible for receiving, processing, storing, and transmitting information from inside and outside of an animal's organisms. The NS is a complex collection of nerves and specialized excitable cells known as neurons, which transmit signals between different parts of the body. The connections among neurons, and between the neurons and the body, are made by means of synapses. Putting it simply, a synapse is a small gap between two neurons. By transmitting signals between different parts of the body, the NS controls both the voluntary and involuntary body actions. It is present, with different levels of complexity, in most multicellular animals [KOC 99]. Sponges are the only multicellular animal to lacks an NS, even though they have some homologous structures that play the role of the NS, for example, in locomotion.

The NS has three main functions: sensory input, integration of data and motor output. Sensory input is when the body gathers information, or data, by the way of neurons, glia and synapses. Our understanding of the world that surrounds us comes from our sensor cells that are part of the NS. Sight, hearing, touch and pain are all expressions of some sensory input sending information to our brain. It is interesting to note that the brain itself has no pain sensors; once you are inside the head, even if you cut a part of the brain, there is no pain. If we think logically, and from a strictly evolutive point of view, it makes sense. If a predator manages to pass through the cranium to reach the brain, it is already too late for the purpose of expressing any pain reaction. The third function of the brain, the integration of the data, is about making sense of all the acquired data, planning reactions and taking actions, which can involve controlling the body, i.e. motor output.

3.1. Nervous system hierarchy

In most animals, the NS consists of two parts: central and peripheral. In vertebrates, the brain and the spinal cord compose the central nervous system (CNS). The peripheral nervous system can be further divided into the somatic nervous system and autonomic nervous system [PAS 12].

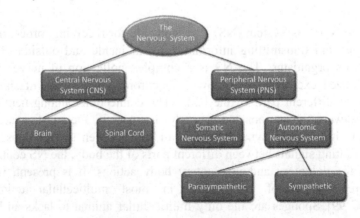

Figure 3.1. *Hierarchy of the nervous system*

3.1.1. *Central nervous system*

The CNS, composed of the brain and spinal cord, is the largest part of the NS. It compiles the information received from all the parties of the body and coordinate the body activities. In some sense, the CNS works as an information processing system, where, given the available inputs, actions are planned and decisions are made. The CNS contains the higher metabolic part of our organism, i.e. the brain. The brain constitutes about 2% of our body weight, but it uses around 20% of all the oxygen that we breathe and approximately 20% of the blood that is pumped by our heart goes to the brain. The cerebrum, the area responsible for the control of the voluntary muscles and the reasoning, represents up to 85% of the brain's weight.

3.1.2. *Peripheral nervous system*

The peripheral nervous system is composed of somatic and autonomic nervous systems. The somatic nervous system is responsible for somatosensation and conscious/purposeful actions, e.g. the voluntary muscle control. The autonomic nervous system is responsible for the involuntary, or vegetative, processes. It regulates systems that are unconsciously carried out to keep our body alive and well, e.g. heartbeat, respiration and digestion. The autonomic nervous system can be further divided into sympathetic and parasympathetic, responsible for the opposing processes of arousal and relaxation. These two subdivisions work without conscious effort and have similar nerve pathways. However, their effects on target tissues present opposite effects [JAN 89]. We can perceive the action of the sympathetic nervous system when, for example, we get excited or are in danger and our heartbeat increases. The sympathetic nervous system is related to the body's *fight-or-flight* responses for possible treats, i.e. it is related to the survival instincts, providing means for the organism to react to possible dangers.

Our body continuously collects sensorial data through neurons, glia and synapses. The neurons operate on excitation or inhibition mode. When in arousal state, the body secretes acetylcholine, which activates the secretion of adrenaline and noradrenaline, which increases the awareness and prepares the body to face danger. The parasympathetic system has the opposite influence over the body, as it is linked to the conservation of energy. It is the system which is responsible for slowing down the heart rate and increasing the intestinal and gland activity. After highly stressful situations, the role of the parasympathetic nervous system is to restore the body's balance as fast as possible, which may lead to overcompensation [WIK 14]. This is why, normally, after an increase in the metabolic reaction caused by the sympathetic system, we have an abnormaly low metabolic reaction (e.g. lower than the normal heartbeat rate) caused by the overreaction of the parasympathetic system in trying to decrease the body's activity.

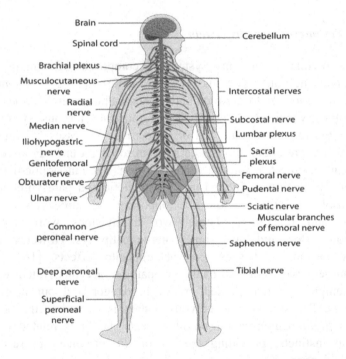

Figure 3.2. *The nervous system diagram (source: "TE-Nervous system diagram" The Emirr, Licensed under CC BY 3.0 via http://commons.wikimedia.org/wiki/ File:TE-Nervous_system_diagram.svg)*

3.2. The neuron

There are three types of neurons in the body: sensory neurons, interneurons and motor neurons. In vertebrates, neurons are found in the brain, spinal cord, nerves and ganglia. At least half our brain cells are neurons. Typically, neurons are composed of three main parts: cell body (soma), dendrites and an axon. The cell body stores the genetic information, cell nucleus and deoxyribonucleic acid (DNA); the axon is the structure that searches for other neurons to create the interconnections. The dendrites are the structures responsible for receiving the signals from other neurons. Figure 3.3 presents the schematic representation of a neuron and its main structures.

The main role of neurons is to process and transmit information, but functionally speaking, they normally act as an adder registering how excited they are and, if excited enough, they trigger a signal to the neighbors. Neurons always send signals with the same intensity; what changes is the frequency in which the signals are triggered. Important signals have a higher frequency-triggering rate than less important signals. Neurons are highly specialized in processing and transmitting cellular signals. As expected, taking into account the diversity of functions performed by them, neurons come in a wide variety of the shapes, sizes and electrochemical properties. For instance, in terms of size, a neuron can vary from 4 to 100 μ in diameter, and the nucleus varies from 3 to 18 μ [WIK 14].

Neurons send signals to other cells as electrochemical waves traveling along the axons (from the Greek ἄξων, (axon), meaning "axis"). These signals activate the release of chemicals, called neurotransmitters, over the neurons junctions, called synapses. Neurons do not touch each other; the synapses, from the Greek συνάπσις (synapsis), meaning "conjunction", are small gaps of approximately 20–40 nm (just for the sake of comparison, a sheet of paper is about 100,000 nm thick). The signals are transferred over this gap via a chemical process, and not an electrical process. Through an impulse, the axon terminals of a neuron release neurotransmitters over the synapses. On the other side, the adjacent membrane (dendrite, muscle or gland cell), with the appropriate chemical receptors, receives the neurotransmitters. This may trigger an electrical impulse across the neighbor neuron cell. The signal starts electrical, becomes chemical and returns to be electrical over the next neuron. The time for neurotransmitter action is between 0.5 and 1 ms. Over the axon, the transmission of the electric pulse occurs by differences of potentials. Instead of passing along the entire axon, the signal jumps from one node of Ranvier to the next.

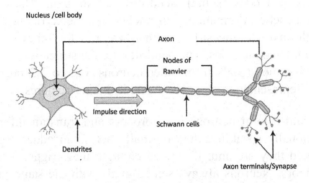

Figure 3.3. *Schematic representation of a neuron, pointing out its main parts*

The soma (cell body) is the central and largest part of the neuron. It contains the nucleus of the cell, and, for this reason, it is the place where most of the protein synthesis occurs. The nucleus ranges from 3 to 18 μm in diameter. The cell body receives the signals from other neurons through the dendrites. Dendrites are cellular extensions connected to the soma. The overall shape and structure is referred to as a dendritic tree. In fact, the term "dendrite" comes from the Greek δένδρον (déndron), meaning "tree".

Our brain is constantly changing; we learn because our synaptic relations change, i.e. our neurons change [KAN 03]. If you are able to remember anything you read from this book, from more than 30 s ago (the time information is stored in the short-term memory [ATK 71]), then your brain has necessarily changed during the reading. If a synapse is used in a consistent, coherent and important way, it tends to be maintained. If 10 years from now you still remember something, anything, from this book, this means that the connections were stable and the change, in a 5 s reading of a single phrase, was preserved all this time. The neuron changes constantly; it grows, and when growing it samples the environment. For example, in case it reaches another neuron, which presents a repulsive cue, the axon abandons that direction and tries another. The search is more or less random, but there are moments where the reorientation occurs, given attractors and repellers. The axon tends to grow in the direction of attractors, and away from repulse signalings [TES 96].

3.3. The neocortex

Many conjectures exist about the speed capacity and functionality of the brain. What we know is that the neocortex is what defines what we are. The neocortex, Latin for "new bark" or "new cover", is the, relatively, new structure that covers the mammalian brain, particularly in humans [MOL 14] (new in evolutionary terms). It is in this part of the brain that all the relations, memories, superior thinking and reasoning are supposed to happen [HAW 05]. More than two-thirds of the neocortex are folded into grooves. The grooves increase the brain's surface area, facilitating inclusion of a larger number of neurons. In humans, the thickness of the cortex varies from 2.3 to 2.8 mm, and is organized into six layers, numerated from I to VI, counting from the most external to the most internal one. Just for the sake of comparison, it is more or less the size of a stack of six playing cards.

The division was made primarily considering their visual distinction, but each cortical layer contains different neuron shapes, sizes and densities as well as different organizations [SWE 06]. Layer I, called molecular layer, is basically a mat of axons, which has very few neurons. Both layer II, the external granular layer, and layer III, the external pyramidal layer, are composed of packet pyramidal neuron cells. This kind of neurons are the most common neurons in the neocortex; for example, eight out of 10 neocortex neurons are pyramidal neurons. Their name is linked to the fact that their cell bodies are shaped roughly like pyramids. Layer II is distinguished from layer III by presenting neurons with slightly smaller cell bodies. Arguably, this size difference is only given to the proximity of layers I and II, and a large part of the ramifications of layer II are, in fact, used to compose layer I. Layer IV predominantly consists of a large number of star-shaped and densely packed cells (also known as "granular" cells), with a noticeable absence of pyramidal cells. Layer V, the internal pyramidal layer, is composed of regular pyramidal cells and some larger than average pyramidal cells. Layer VI, the multiform or fusiform layer, is characterized by the large variation in neuron types.

The relation between brain weight and cortical thickness is approximately logarithmic [NIE 98]. Humans are smarter than other mammals; it is not because they have some special "clever" cells or because the layers of their cortex are thicker. Our reasoning capacity is most probably related to the fact that our cortex, relative to body size, covers a larger area than the average of other mammals. The human neocortex has about 22.8 billion neurons [PAK 97] into a total surface area of approximately 2,320 cm^2 [PET 84], which would fit an area of 50 × 50 cm^2. The basic composition and organization of the cortex of different species is similar.

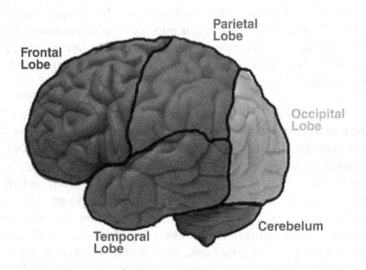

Figure 3.4. *Main regions of the brain*

The cortex is divided into specialized zones, similar to a continental map. Each one of the cortex areas is responsible for one specific task. The frontal lobe is responsible for motor coordination and higher cognitive skills (e.g. problem solving, thinking, planning and organizing). The frontal lobe also contains aspects linked to personality and emotions. The parietal lobe is involved in sensory processes, including orientation, attention, mathematical reasoning and language. The occipital lobe contains the primary visual cortex and helps in the processing of visual information, including shapes and colors. The temporal lobe contains the primary auditory cortex and helps in the processing of auditory information and integrating information from other senses.

There is evidence that the temporal lobe plays a major role in the fixation of memories. We have two temporal lobes; if one presents a problem, it is not extremely serious, as the memory functioning is taken care by the other temporal lobe. However, there is at least one attested clinical case where, after losing both temporal lobes, plus the hippocampus, an individual develops problems with the fixation of short-term memories, even though previous long-term memories remain intact [SCO 57]. Henry Molaison, or patient H.M. as he was called in the numerous articles written about him, went through an experimental procedure in the 1950s in which he had both his temporal lobes and the hippocampus removed [CAR 08]. Neither his long-term not short-term memory was really affected. However, he lost the capacity to fix new memories, i.e. transform the short-term memories into long-term memories. Another interesting fact is that, apparently, his procedural memory was also intact. The procedural memory holds the memories related to our motor skills. If a motor skill-related task was presented to him, he progressed performing the task. The first time he would take more time to perform the task, but over the time, though repetitive sections, the time to perform the task would decrease, the same way as with any regular person. Even if H.M. could not remember of doing it previously, it showed that the transfer for the long-term memory was working, at least to motor skills-related tasks.

In terms of relative area, the frontal lobe occupies around 41% of the neocortex area, the temporal lobe occupies about 22%, the parietal lobe occupies about 19% and the occipital lobe occupies about 18%. However, the divisions of the areas and their specializations, even inside these areas, are not proportional. For example, if we consider the sensorial and motor parts of the cortex, and we relate them to the areas of our body they control, we can see that the distribution is unequal. If our body was proportional to the sensorial areas of our brain, we would have a quite deformed body [SCH 93]. The homunculus proposed by Penfield and Rasmussen [PEN 50] is a representation that tries to show this disparity between neocortex cartography and the parts of the body they are concerned with.

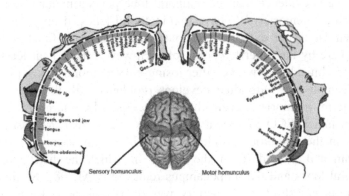

Figure 3.5. *Schematic representation of the brain regions and the body areas they control, adapted from the homunculus in [PEN 50]*

3.4. Speed and capacity

The speed of the propagation of the signals varies greatly; an impulse may travel from 1 to 100 m/s, depending on the type of the neuron; it also takes 0.5 ms for a signal to cross the 20 nm synapse gap [SIR 08]. The neuron needs about 1 ms to fire an impulse and return to its resting level [COO 89] and, therefore, to be able to fire other impulses. This means that, theoretically, we would have the possibility to transfer an impulse every 1.5 ms, i.e. about 660 times per second. Even though this varies considerably, it is considered a typical neuron-firing rate to be once at each 5 ms (about 200 times per second, i.e. 200 Hz) [FRE 10]. Moreover, on average, one neuron is activated by 1,000 neurons, which activates an other 1,000 neurons [HUY 01]. With these values, we can make some interesting estimations about the human brain and its capacity. However, it is important to highlight that these calculations are based on estimations and averages; therefore, by definition, they are intrinsically wrong. Nevertheless, these quantitative analyses can give us an idea of the power the mechanisms the brain dispose of and how massive the processes performed at each moment are.

Lets consider that all the brain computations are made in our cortex and in it there are ~22.8 billion neurons [PAK 97]. If we assume that each exchange between synapses is one bit of information, 0/1, firing 22.8 billion neurons around 200 times per second and spreading the firing over 1,000 synapses each time, in terms of bits of transmission over the cortex at each second, we have 4,560,000,000,000,000 bits (4.5 millions of billions), i.e. ~570 terabytes of information exchanged at each second. This is a huge simplification, as the intensity of the chemical reaction has an impact, thus changing the value of the information, i.e. the relations are not 0/1. Taking this into account, the value is probably much higher.

According to Thorpe *et al.* [THO 96], the time to recognize an image is believed to be as fast as 150 ms, which means that the chain of neurons (the maximum path if we prefer) involved in the search for this information cannot be higher than 100 neurons. Suppose that a neuron needs 1 ms to recharge and 0.5 ms for the signal to cross the synapse. If we consider that the chain contains 100 neurons, and if each neuron activates 1,000 other neurons, then we could estimate the number of synapses involved in a visual memory recall.

As a simplification, let us assume that the process starts with one single neuron. The signal propagation can then be modeled as a geometric progression in which the first term is 1 and the common ratio is 1,000. For a 100-level geometric progression, we could have as much as 10^{297} synapses involved in the image recognition process, which would be bigger than the expected number of synapses in the whole cortex. Considering our numbers, we reach the conclusion that we have ~22 × 10^{11} synapses (22.8 billions × 1,000 synapses = 22 × 10^{11}). Connors and Long [CON 04] on their part estimated the brain to have around 10^{15} synapses, still much lower than the 10^{297}. Just for the sake of the comparison, the number of stars in the galaxy is forecast to be around 4 × 10^{11} [CAI 13].

Considering these numbers, we can infer that each synapse participates more than once in the process. In fact, it is interesting to note that the same principle is valid over different natural systems. In general, the important factor is not related to the elements that compose the system, as much as the way they are organized [DIT 01]. Our brain is not an exception: the synapses or the neurons are not as important as the way they are organized and interconnected. The cortex contains hundreds of thousands of neurons and we are the expression of how these neurons are interconnected.

3.5. Artificial neural networks

As a biological neuronal network, artificial neural networks (ANNs) are made up of a series of artificial neurons. The number of neurons depends on the task to be fulfilled; it may vary from one to several thousands. Each artificial neuron is a mathematical simplification of a biological neuron. The mathematical model that serves as the basis for the ANN was proposed by McCulloch and Pitts [MCC 43] in 1943. At the time they called their technique threshold logic.

The most attractive characteristic of ANNs is their capacity to learn and solving problems that are either too complex for standard algorithms or even that there are no algorithms available to solve. As the ANNs are inspired by our NS, they excel in the same problem-solving tasks we do. Computers can perform calculations in an incredible speed; within seconds they can do precise mathematical inferences that would take someone an entire life to perform. However, on the other hand, there is a whole set of tasks that are simple for us but extremely difficult for computers to fulfill, e.g. to recognize whether a photo is of a dog or a cat. Moreover, the ANN is a non-parametric method, i.e. it makes no assumptions regarding the statistics of the inputs. For this reason, it can be viewed as a generic problem-solving method; we can use it in the resolution of a broad range of problems. ANNs have a remarkable capacity of extracting meaningful information from complex or imprecise data, detecting patterns and trends that are too complex for traditional algorithms, and even humans, to distinguish.

Based on an initial training experience, ANNs are capable of self-organization and self-adaptation. They autonomously create their own model of the problem and learn how to treat the task at hand. After training, an ANN can be seen as a specialist in that specific domain. Given the gathered information, it can infer relations and make projections even when there is incomplete input information available. Moreover, it can take the most of parallelization, specially when we consider hardware-specific implementations [WID 61, NUG 02]. All these are remarkable characteristics and some of the main reasons why ANN is a popular and wide spread technique.

3.5.1. *The perceptron*

The perceptron [ROS 58] is the abstraction of a real neuron and the basic component of an ANN. The perceptron is a simplified version, a model, of a real neuron. How efficient a perceptron is when compared to the natural neuron is still an unanswered question; however, the efficiency of the perceptron in solving complex problems is indisputable. A perceptron has a series of inputs, stimulus, each one having a relative importance, i.e. weight. At the beginning, the weights are traditionally initialized with some small random values. Their values are adjusted, or updated, during the training period. The role of the weights is to amplify or decrease the importance of a given input value over the perceptron's output. All the inputs, multiplied by their respective weights, plus a bias also multiplied by its respective weight, give us a value called activation value. The perceptron activation value is the measure of how excited a perceptron is. The result of the activation value serves as input to an activation function, responsible for deciding the perceptron's output. The bias is a constant term, normally 1, which does not depend on any input value. The function of the bias is to provide a minimum threshold value to the neuron. As it is a constant value, it alters the position of the decision boundary. The bias has an important role in the learning process, mainly in multilayer ANNs, as it provides a way to dynamically change the relative importance/impact of the perceptron.

The activation function can be any function that, given the activation value, can decide about the output of the perceptron. The most common ones include step functions and sigmoid functions. A step function, used in the original perceptron, is a function in which the output is based on a threshold; for values below the threshold, the output is 1, and for values above the threshold the value is 0. For example, if the input is below 0.5, the output value is 0, and if the input is above 0.5, the output is 1. This kind of activation function is useful when we want to classify inputs into two groups. They can also be used in multilayer perceptron networks, where the output of one perceptron is linked to the input of others, which enables for complex problem solving. Even if the output of each perceptron is a simple 0/1.

A sigmoid, or logistic, function has the same role as step functions, but with additional uncertainty region, i.e. the decision of the output does not rely on a single cutting point value. In this respect, sigmoid functions are closer to biological neurons than step functions. The sigmoid function and its relation with the perceptron are shown in Figure 3.6.

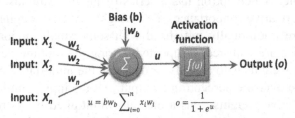

Figure 3.6. *Schematic representation of the classical perceptron with the sigmoid activation function explained*

Even if a single perceptron is already a powerful tool, capable of solving a large series of problems, it has some significant limitations.

To be decidable by a single neuron, the answer domain needs to be linearly separable. Minsky and Papert [MIN 69] showed that a single perceptron was incapable, for example, to decide the output of a simple XOR function. If we represent the inputs and outputs of an OR function in a graph (see Figure 3.7(a)), we can see that the result can be separated with a single straight line. Problems that have this kind of characteristic on their solution space are called "linearly separable". This is the kind of problem that can be modeled with a single perceptron. On the other hand, one cannot separate the values of an XOR function with a single line (see Figure 3.7(b)). Werbos [WER 74] proposed the backpropagation algorithm, which is the first efficient solution to this limitation of the perceptron.

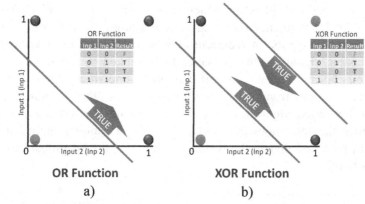

Figure 3.7. *OR and XOR figures create a table to show the points*

3.5.2. *Interconnecting perceptrons*

Calling a single perceptron an artificial neuronal *network* does not make much sense. By definition, a network should have at least two interconnected perceptrons. However, the practical advantage of linking perceptrons together is that it permits us to solve a large range of problems, notably nonlinearly separable problems. Different researchers have proposed to interconnect the perceptrons in different ways, however, two of the most common and efficient ways are feed-forwarding and backpropagation. Feed-forward is the most simple and, probably, the most common way to interconnect perceptrons.

A feed-forward ANN contains only forwarding paths, perceptrons organized into layers with each layer receiving input from the previous layer and outputting the results to the next layer. There is no feedback path, i.e. no signal is transmitted to a perceptron located in a previous layer. The nodes may be fully, or sparsely, connected; a network in which the output of a perceptron is not forwarded to all perceptrons in the next level is known as a sparsely, or non-fully, connected ANN. The percentage of available connections that are utilized is known as the connectivity of the network.

The second type of ANN, which is also quite common, is recurrent ANNs, or backpropagation when we are talking about feeding back the evaluation errors. In recurrent networks, we have feedback paths that enable signals from one layer to be linked to the inputs of a previous layer. A fully recurrent network is one where every neuron receives input from all other neurons in the system. The backpropagation algorithm is a way to, during the training period, consider the total network output error for a training input/output pattern pair, and update the weights to improve the hit probability of the network.

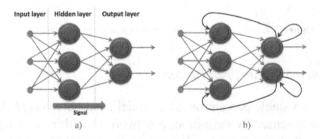

Figure 3.8. *The standard architecture for ANNs with three layers: an input layer, a hidden layer and an output layer in a) a feed-forwarding and b) a recurrent representation*

Figure 3.8(b) shows a non-fully recurrent network, i.e. not all notes point to all others. It is a simple three-layer network: an input, an output and a hidden layer. We call hidden perceptron any perceptron that is neither in the input layer nor in the output layer, i.e. they are hidden from the view. They work more or less as a black box to anyone that is handling the ANN. The use of hidden layers of perceptrons grants the system a larger expressivity and flexibility in threatening problems. However, this flexibility has a price, the increase in the complexity of the ANN.

When designing an ANN with hidden layers, two basic questions need to be answered. How many hidden layers should be created, and how many perceptrons should each layer have? It is true that people have used deep neural networks (DNNs) [ZHA 14], i.e. ANNs that have two and more hidden layers, and it is also true that for some difficult problems, such as speech recognition, it can be interesting to have DNNs [DAH 12]. However, in practice, few problems require more than one hidden layer, and there is no theoretical reason to use more than two hidden layers [HEA 08]. If we do not use hidden layers, we are limited to solve only linearly separable problems. One hidden layer lets us perform any nonlinear mapping [HAY 99]. With two hidden layers, we can represent the arbitrary decision boundaries and can approximate any smooth mapping to any accuracy [HEA 08].

After deciding the number of hidden layers, the next and most subtle problem is to decide the number of perceptrons each one of these layers will have. Using too few perceptrons may result in underfitting, i.e. the number of perceptrons is not sufficient to correctly detect the signals in a complex data set. Using too many neurons in the hidden layers can result in several other problems. The first problem is overfitting, i.e. the network gets so specialized, and linked to the training data set, that it may be useless, or perform poorly, when applied on real data sets. The second problem that may occur is that the time to train the ANN may increase significantly with the number of hidden layers.

Even though the fine-tuning of the ANN is ultimately a trial-and-error process, i.e. different configurations may need to be tested to see which one adapts better to the problem, Heaton [HEA 08] lists some heuristics to help in the first guesses to know how many perceptrons the hidden layers should have:

– set the number of hidden perceptrons to a value that is between the size of the input layer and the size of the output layer;

– set the number of hidden neurons to be 2/3 of the size of the input layer, plus the size of the output layer;

– the number of hidden perceptrons should be less than twice the size of the input layer.

3.5.3. *Learning process*

In the beginning, the ANN is generic; it has no defined objective. It is like an empty box that we can fill with whatever we want. It becomes the toy's box only after we fill it with toys; if, on the other hand, we decide to fill it with socks, it would become the socks' box. To be useful in solving a specific problem, an ANN needs to pass through a process of training. Over this phase, the weights are changed and the ANN converges to a model adapted to treat the task it needs to perform. We say that the non-trained state is generic because by changing the weights we can get different models of decision-making, i.e. for the same inputs we may have radically different outputs. What teaches the ANN if the output is the one we expect/want is the training/learning process.

We can find various ways to perform the training over the literature, each one with its own advantages and disadvantages. It is important to note that the training process of an ANN is an interactive process, where the weights constantly change at each interaction. In fact, the objective of the learning phase is to converge to a set of weights that, when applied to the ANN, will, hopefully, map any input to the correct output. The three main learning paradigms on artificial intelligence, and which can also be applied to ANNs, are supervised, unsupervised and reinforced learning.

Supervised learning is a controlled process. There exists a set of inputs to which the outputs are known. These pairs of input and expected outputs are presented to the ANN that evaluates the inputs, and the result of the ANN is compared with the expected output. Then, we calculate the error between the actual output and the expected output. The error is then used to make corrections to the weights trying always to decrease the error. A common method is to use genetic algorithms (GAs) to perform an exploration over the weights state space. The genome is coded as the set of weights, and the fitness function, in general, is the error. The objective of the GA is to decrease the error as much as possible.

In unsupervised learning, the ANN is presented only with the set of inputs, but not with the outputs. It is the responsibility of the ANN to find a pattern in the inputs without any help. At each training iteration, the trainer provides the input to the network, and the network produces a result. This result is put into the cost function, and the total cost is used to update the weights. Weights are continually updated until the system output produces a minimal cost.

Reinforcement learning is close to supervised learning in the sense that some feedback is given; however, instead of trying to reach a target output, rewards are given to the network if it relates correctly to the input and output. The main goal of the ANN in this case is to maximize future rewards. The roots of this type of learning mechanism are linked to the evolutionism, where, for example, a given predator remembers the previously used successful tactics in finding food, i.e. the reward.

3.5.4. *The backpropagation algorithm*

Now it should be clear that the efficiency of the ANN lies in the correct setting of the weights. One popular way to choose/update these weights to reach the correct result at the output layer is the backpropagation algorithm [ROJ 96]. Broadly speaking, the algorithm consists of feeding the derivative of the total network output error backwards for a training input/output pattern pair. This value is then used to update the weights with the intention of decreasing the error.

The weight changes are calculated by using the gradient descent method [BOT 98]. This means that we follow the steepest path on the error function to try to minimize it. In other words, what we need to do is to take the error at the output neurons (desired value – reached value) and multiply it by the gradient of the sigmoid activation function. If the difference is positive, we need to move up the gradient of the activation function, and if it is negative, then we need to move down the gradient of the activation function. To be able to do so, the activation function needs to be differentiable, i.e. a function where the derivative exists at each point in its domain. A number of functions fulfill this requirement, and could be used, but one of the most used functions is the sigmoid function [TVE 01]:

$$O_j = \frac{1}{1+e^{-Act_j}} \qquad [3.1]$$

where Act_j is the activation value for perceptron j given by:

$$Act_j = \sum_{i=0}^{n}(x_i * w_{ij}) + b * w_{bj} \qquad [3.2]$$

where n is the number of inputs, w_{ij} is the weight applied to input x_i, b is the bias, always 1, and x_{bj} is the weight applied to the bias of perceptron j.

In the output layer, considering a given perceptron k, the error is given by the difference between the desired value (d_k) minus the actual output for that perceptron O_k, we can express this difference by Δ_k, where $\Delta_k = d_k - O_k$. The actual error signal to be backpropagated is given by δ_k, where:

$$\delta_k = \Delta_k * O_k(1 - O_k) = O_k(d_k - O_k)(1 - O_k) \qquad [3.3]$$

where the term $O_k(1 - O_k)$ is the derivative of the sigmoid function [3.1]. If the difference is positive, we need to move up the gradient of the activation function, and if it is negative, then we need to move down the gradient of the activation function.

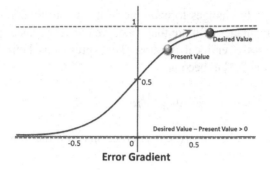

Figure 3.9. *Error gradient representation*

With the delta rule, the change in the weight connecting input node *j* and output node *k* is proportional to the error at node *k* multiplied by the activation of node *j*. The new adjusted weight, $w_{jk(new)}$, between the output perceptron *k* and the input *j*, based on the present w_{jk}, is given by:

$$\Delta w_{jk} = l_r \delta_k x_{jk} \qquad\qquad [3.4]$$

$$w_{jk(new)} = w_{jk} + \Delta w_{jk}, \qquad\qquad [3.5]$$

where Δw_{jk}, is the change in the weight between perceptron *k* and input *j*, x_{jk} is the value of the input *j* and l_r is the learning rate. The learning rate is basically how much we move over the error gradient. Large l_r means larger steps and small l_r means smaller steps. The setting of the learning rate needs to be carefully planned. Excessively large steps may mean that we will go over the most adequate value, not ever being able to really find it. Extremely small steps, on the other hand, may mean that the convergence to the best point may take a long time. A typical value for the learning rate is 0.01; however, it is also possible to have a dynamical learning rate. In the beginning of the training, we have a large l_r that decreases as we converge to the best weight. This approach makes sense since the initial weight values are basically chosen at random, therefore probably far from the optimal set. Another way to express dynamicity is the application of a *momentum* (*μ*) correction factor [SIV 03]. This momentum normally, linked to the number of interactions, fosters the weight changes to continue in the same direction with larger steps. The advantages of it are that, first, we converge faster and, second, we tend to pass over local minima. To express this behavior, we can change formula [3.4] to become:

$$\Delta w_{jk}^{i} = l_r \delta_k x_{jk} + \Delta w_{jk}^{(i-1)} \mu \qquad [3.6]$$

where i is the number of present interactions, and the momentum has a value between 0 and 1.

The calculation of the error gradient on the hidden and output layers is different. In the hidden layer, the error gradient is based on the output layer's error gradient (backpropagation). Therefore, for the hidden layer, the error gradient for each hidden neuron is the gradient of the activation function multiplied by the weighted sum of the errors at the output layer originated from that neuron. Well, the mathematical intuition is perhaps not exactly simple, but, believe or not, the application of these concepts is much simpler than it looks like. The error signal for node j in the hidden layer can be calculated as:

$$\delta_k = O_k(d_k - O_k) \sum w_{jk} \delta_k \qquad [3.7]$$

where the sum term adds the weighted error signal for all nodes, k, in the output layer. Formulas [3.6] and [3.4] basically remain the same, only that we now also consider the inputs of the hidden nodes.

We end the learning process when the global error is zero, which is hardly the case. The most common thing is to consider a threshold below which we can consider the error as negligible. We can use the following equation to calculate the error function, E, for all tested patterns (tuple inputs/expected outputs):

$$E = \tfrac{1}{2} \sum (\sum (d_k - O_k)^2) \qquad [3.8]$$

3.6. Applications on networks

ANNs are probably the most popular artificial intelligence method in computer networks. Virtually all network-related areas have at least one work proposing to approach problems from an ANN point of view. A non-exhaustive list includes intrusion detection systems [RYA 97, CAN 98, GHO 98], fault detection [JAG 14], energy efficiency [ENA 10], routing [RAU 88, ALI 93, BAR 06], medium access control [KUL 09] and handover [NAS 07]. However, perhaps the most ambitious project is the "Autonomic Computing" initiative, which was launched by IBM in 2001 [KEP 03]. This initiative was motivated by the need to find better ways of managing the increasing complexity of computer systems, especially network ones [STE 03]. Autonomic computing was meant to set the foundations for a new generation of computer systems able to self-manage. When IBM's senior vice president of research, Paul Horn, introduced the autonomic computing idea in 2001, it was clear that the inspirations for the new initiative came from the mammalian autonomic nervous system. "The autonomic nervous system governs our heart rate and body temperature, thus freeing our conscious brain from the burden of dealing with these and many other low-level, yet vital, functions" [KEP 03]. Even though the idea was not entirely new, i.e. some people had already worked on this kind of approach before [HUE 08], IBM popularized it and, *de facto*, started a whole new movement with the purpose of creating computing systems that could self-manage.

3.6.1. *ANN in intrusion detection systems*

A simple and effective initiative for using neural networks in intrusion detection is neural network intrusion detector (NNID). In their work [RYA 97], Ryan *et al.* proposed to use a backpropagation neural network to distinguish attackers from the regular users of the system. The proposed approach consists of collecting the commands issued by the users and recognizing the usage pattern of each user. NNID is able to recognize when an attacker is impersonating a rightful user. The authors argue that each user has a particular way of using the machine. It can be characterized by the frequency of commands and the selection of tools people use; for example, some Unix users are fond of *Vi* as a text editor, while others would prefer to use *Emacs*.

The method essentially consists of three steps:

1) collecting the training data: the number of times each user issued each command organized into usage vectors;

2) training: letting the neural network determine the usage patterns for each one of the users of the machine, having as input the usage vectors;

3) performance: using the NNID to identify the users of the machine. If the user is not identified, or no clear indication of the user is reached, an anomaly alarm is raised.

In the evaluation of NNID, the authors used a Unix-like machine with 10 regular users, considering the 100 most common commands issued over the machine. A total of 11 days of logs were collected, from which eight random days were used for training the neural network, and the remaining three days were used for evaluating the results. The actual usage of each command was represented by 11 nonlinearly spaced interval levels, where the first interval means no use of the command for the given user, the second interval means that the command was used at most two times, and so forth until the 11th level, which means that the command was used more than 500 times. The user usage distribution vector was used as input for the NNID. The input neural network layer consisted of 100 units representing the user vector; the hidden layer was composed of 30 nodes and the output layer was composed of 10 nodes, one for each user.

In its simplest form, the system achieved 93% of correct user identification and false alarm rate of 7%, a remarkable rate considering a low complexity of the approach. It is also interesting to note that the errors were always linked to the same user. It is found that the user was an infrequent user, and the training data set counted only with three log days for him. This leads to a less robust representation of this specific over the NNID.

Lippman and Cunningham [LIP 00] used a mix of neural networks and word count to search for attacks on the traffic transmitted on the network. It is a more interesting idea since the attacks can be detected online while they are happening, and countermeasures can be taken to avoid the attack. Approaches based on the logs, such as NNID, can help us detect whether an attack has happened, but by that point it is too late to act since the attacker will have already accessed the machine.

In a traffic monitoring kind of attack detection system, a *sniffer* collects the traffic data and the content of the packets is analyzed in search of a predefined list of words that may be indicators for attacks. The main objective of the neural network on the proposed architecture is twofold: first, it improves the efficiency of traditional keyword searching methods and, second, it decreases the number of false alarms raised. Keyword searching techniques are quite efficient, but have a large number of false alarms [LIP 00], sometimes 100 per day.

Lippman and Cunningham's technique uses a multilayer perceptron with no hidden layers. The evaluation of the technique was done over the DARPA Intrusion Detection Data Set [CUN 99], with only 20 training cycles and 30 keywords. The used gradient descent step size was 0.1. These keywords were a mix of line commands and program answer strings. The neural network automatically classified the words attributing weights to them, some receiving positive weights and some receiving negative weights. This can be used to define which words are common in attacks and which words are more common in normal sessions.

More information about the use of computational intelligent approaches applied to intrusion detection systems can be found in [WU 10].

3.6.2. *Fault detection*

Fault detection represents the first step in the fault diagnosis tool chain. It consists of the identification of whether some abnormal behavior has occurred and is related to the reliability of the system and its results. Jager *et al.* [JAG 14] have studied the problem from the viewpoint of sensor systems. The technique uses a data-driven fault identification approach. Each sensor and the type of data can be acquired from the way it is classified. The types of faults observed are also identified and their characteristics modeled to be used as the fitness function.

A subset of possible signal input features is carefully chosen among the ones that are computable efficiently, uncorrelated with other features, independent of external influences, and characterized by high differences between features and small internal differences. The intention is to choose inputs that can guarantee the applicability of the approach on a wide range of sensor types. The features extraction needs to make the data acquiring and the fault detection disjoint. In this case, the inputs are a time series, i.e. a constant flow, of signals generated by different sensors.

The sets of faults observed are: outliers, constant offset, constant noise and stuck-at-zero. The ANN covers all these fault types. The features used as inputs for the neuronal network are the following mathematical transformations over the input namely: mean, standard deviation, deviation, signal-to-noise ratio and correlation coefficient. The inputs for the ANN are six flows with different mathematical transformations applied to the raw input data. The implemented ANN in this case is a feed-forward one. The number of hidden layers and the number of nodes on each hidden layer are a parameter of the problem. They evaluate different configurations to choose the ones that present the best results.

Christensen et al. [CHR 08] use time-delay neural networks (TDNNs) to detect faults in autonomous robot systems. Training data are collected over a series of robot's normal operations. The main idea is that hardware faults change the flow of sensory data and, as a result, the reaction of the control program. The TDNN is used to detect such changes. TDNNs are feed-forward networks that allow reasoning based on time-varying inputs without the use of recurrent connections. The TDNNs used in this study are normal multilayer perceptrons for which the inputs are taken from multiple, equally spaced points in a delay-line of past observations. In the evaluation, the authors used one hidden layer of five neurons, fully connected to the output layer, and an input layer with 10 input groups from different moments in time. The advantage of this approach is that it is adaptable to classify data based on current and past observations, i.e. it permits the observation of parameters that, over time, lead to a fault recognition. Standard backpropagation is used in the learning phase; the TDNN is trained to discriminate between robots normal and faulty operation.

The problem with TDNN is the static size of the time window; some faults may not fit well to a statically defined time window. One way to overcome this disadvantage is the use of locally recurrent neural networks (LRNNs) [PRZ 08]. The study object in this case was a system with three water tanks. Information on different sensors was available to detect faults inside the model. Dynamic neurons are defined with a specific linear dynamic system that retransmits the output back to the input. This recurrent connection enables the LRNN to deal with time series dynamically.

3.6.3. *Routing*

Rauch and Winarske [RAU 88] present one of the first attempts to apply artificial neuronal networks on routing. At the time, they evaluate their method in a "large" 16-node network. The authors argue that traditional sequential routing algorithms are not efficient in determining optimal routes when applied to large communication systems. The implemented method looks for the best one-link routes, two-link routes and so on, up to a maximum of five link routes from each node. The method is interactive and converges after approximately 200 interactions.

Initially, the ANN has no knowledge of the actual node-to-node traffic; therefore, the initial loss function w_{ij} is chosen (somewhat arbitrarily) to be equal to the inverse of the node-to-node capacity C_{ij}:

$$w_{ij} = \frac{1}{c_{i,j}}.$$ [3.9]

In the next iterations, the actual traffic matrix, determined in the previous step, is used to determine the loss function, and determines w_{ij}. Then, the optimal routing and allocation procedure is repeated to obtain a new actual traffic matrix. This procedure is repeated, for each origin and destination pair, for several iterations to evaluate convergence.

Rauch and Winarske's [RAU 88] method has a great merit for being one of the first attempts in the field. However, the organization of the ANN is fixed to the size of the network and the number of links each node represents. Even if they consider the possibility of downlinks, its implementation in modern dynamical communication networks would be impractical.

Sensor intelligence routing (SIR) [BAR 06] is a routing algorithm that uses ANNs to efficiently control the routes in public utility services sensor networks. Public utility services meters are nowadays starting to be equipped with networking capacities, which means that the reading of the meters can be done automatically. SIR is a proposition of a Quality of Service (QoS)-driven routing algorithm to manage the flow of information from the meters to a base station. Each node has an ANN to manage the routes for the data flow. SIR is based on Dijkstra's algorithm; it finds the minimum cost paths from the base station to every node. The weight parameter (w_{ij}) for each link between nodes is determined in each node by a self-organizing map (SOM), based on QoS parameters latency, error rate, duty cycle and throughput.

The QoS is determined locally, instead of globally. Each node sends a ping packet to each of its neighbors to learn the specific quality of the link. After acquiring the characteristics of the link, the node requests to the ANN the distance to the base station passing through this path. The type of ANN used is called SOM [KOH 82]. The organization of the SOM implementation consists of a first layer with four neurons and a second layer with 12 neurons in a 3 × 4 matrix. SOM inputs are the values of QoS acquired with the ping packets: latency, throughput, error rate and duty-cycle. The samples presented as inputs to the SOM form groups with similar characteristics. In the resulting clusters map formed, every cluster relates to a specific QoS and to which a neuron of the output layer is assigned.

In the learning phase, executed in a central processing unit, the neurons from the second layer compete for the privilege of learning from each other, while the correct answer is not known. In the execution phase, the neurons' input weights are fixed. After the winning neuron is elected, the node uses the output function to assign a QoS estimation. The highest assignment represents the scenario in which the link measured has the worst predicted QoS, while the lowest assignment represents the best predicted QoS. SIR presents a good performance but the cost in terms of ping messages, to keep the weights updated, is relatively high.

In [SHA 08], Shahbazi *et al.* also proposed an approach based on the use of SOM. The method creates a SOM which describes a minimum power routing model for wireless sensor networks (WSNs). The objective is to increase the network lifetime by optimizing the routing according to the amount of energy of each node in the network. The work builds up a power optimization and energy conservation methods based on each node role. The algorithm takes into account the importance of the node's position for the overall network routes optimization. Shahbazi *et al.* define network life time (NLT) as being the sum of the nodes importance in routing at time *t* and the amount of energy consumption of the node for the routing. A self-organizing neural network is used to decide whether it will participate in the routing of the packet or drop it. As soon as a packet arrives, a feature vector is extracted and this vector is sent to the self-organizing ANN of that node as input. The objective of the ANN is to maximize the NLT parameter. Nodes compete for the right to forward the packets; the nodes that loose the competition drop the received data packets.

The random neural networks with reinforcement learning (RNNRL) [GEL 01] algorithm uses intelligent packets instead of intelligent nodes to control the routing of packets in the cognitive packet network (CPN). CPN is a connectionless packet switching network with intelligent capabilities added to the packets rather than the nodes. Smart packets transit on the network and each node executes the code inside this packet. These smart, or cognitive, packets are sent in the network to discover new routes from a source to a destination, taking into account different QoS metrics. The smart packet routing decisions are either taken randomly or are based on the RNNRL algorithm. In the RNNRL, for each destination and QoS class, a node maintains a recurrent random neural network (RNN). Each neuron in an RNN corresponds to a neighbor of the node. The weights and the internal expectation of the reward are updated whenever a new reward reaches the node. The weights of the RNN are reinforced, or weakened, depending on how they have contributed to the success of the QoS goal. The reinforcement value is QoS metric of the communication system such as path delay and packet loss. The reinforcement is also a function of the network node. For example, if the reinforcement is associated with the end-to-end delay in a source-destination pair, then the reward is the inverse value of the delay from the particular node to the destination. Random decisions contribute to the exploration of the network, while the reinforcement learning-based decisions contribute to improve the route for a given QoS metric. The state of the RNN, according to the information gathered by smart packets, is updated via acknowledgment packets.

3.7. Further reading

WELLS R.B., Cortical neurons and circuits: a tutorial introduction, available at http://www.mrc.uidaho.edu/~rwells/techdocs/ Cortical%20Neurons%20and%20Circuits.pdf, April 2005 [accessed on January 2015].

BYRNE J.H., Neuroscience online, the open-access neuroscience electronic textbook, University of Texas Medical School at Houston, available at http://neuroscience.uth.tmc.edu/ [accessed on 29 March 2015].

GERSTNER W., KISTLER W.M., *Spiking Neuron Models, Single Neurons, Populations, Plasticity*, Cambridge University Press, 2002.

For information as artificial neural networks, see the following publications.

KEPHART J.O., CHESS D.M., "The vision of autonomic computing", *IEEE Computer*, vol. 36, no. 1, pp. 41–50, 2003.

JAIN A.K., MAO J., MOHIUDDIN K.M., "Artificial neural networks: a tutorial", *IEEE Computer*, vol. 29, no. 3, pp. 31–44, March 1996.

ROJAS R., *Neural Networks: A Systematic Introduction*, Springer, 1996.

HAYKIN S., *Neural Networks: A Comprehensive Foundation*, Prentice Hall, 1994.

BALDI P., HORNIK K., "Learning in linear neural networks: a survey", *IEEE Transactions on Neural Networks*, vol. 6, no. 4, pp. 837–858, 1995.

TIMOTHEOU S., "The random neural network: a survey", *The Computer Journal*, vol. 53, no. 3, pp. 251–267, 2010.

SUBHA C.P., MALARKAN S., VAITHINATHAN K., "A survey on energy efficient neural network based clustering models in wireless sensor networks", *Proceedings of the 2013 International Conference on Emerging Trends in VLSI, Embedded System, Nano Electronics and Telecommunication System (ICEVENT)*, pp. 1–6, 7–9 January 2013

ENAMI N., MOGHADAM R.A., HAGHIGHAT A., "A survey on application of neural networks in energy conservation of wireless sensor networks", *Recent Trends in Wireless and Mobile Networks, Communications in Computer and Information Science*, vol. 84, pp. 283–294, 2010.

3.8. Bibliography

[ALI 93] ALI M.K.M., KAMOUN F., "Neural networks for shortest path computation and routing in computer networks", *IEEE Transactions on Neural Networks*, vol. 4, no. 6, pp. 941–954, November 1993.

[ATK 71] ATKINSON R.C., SHIFFRIN R.M., *The Control Processes of Short-Term Memory*, Institute for Mathematical Studies in the Social Sciences, Stanford University, 1971.

[BAR 06] BARBANCHO J., LEÓN C., MOLINA J. *et al.*, "Giving neurons to sensors: QoS management in wireless sensors networks", *IEEE Conference on Emerging Technologies and Factory Automation (ETFA'06)*, pp. 594–597, 20–22 September 2006.

[BOT 98] BOTTOU L., "Online Algorithms and Stochastic Approximations", in SAAD D. (ed.), *Online Learning and Neural Networks*, Cambridge University Press, 1998.

[CAI 13] CAIN F., "How many stars are there in the universe?", *Universe Today*, available at http://www.universetoday.com/102630/how-many-stars-are-there-in-the-universe/, 3 June 2013.

[CAN 98] CANNADY J., "Artificial neural networks for misuse detection", *Proceedings of the National Information Systems Security Conference*, pp. 443–456, 1998.

[CAR 08] CAREY B., MOLAISON H., "An unforgettable amnesiac, dies at 82", *The New York Times*, 4 December 2008.

[CHR 08] CHRISTENSE A.L., O'GRADY R., BIRATTARI M. *et al.*, "Fault detection in autonomous robots based on fault injection and learning", *Autonomous Robots*, vol. 24, pp. 49–67, 2008.

[CON 04] CONNORS B.W., LONG M.A., "Electrical synapses in the mammalian brain", *Annual Review of Neuroscience*, vol. 27, pp. 393–418, 2004.

[COO 89] COON D., *Introduction to Psychology: Exploration and Application*, St. Paul, West Publishing Company, 1989.

[CUN 99] CUNNINGHAM R.K., LIPPMANN R.P., FRIED D.J. *et al.*, "Evaluating intrusion detection systems without attacking your friends: the 1998 DARPA intrusion detection evaluation", *Proceedings of ID'99, 3rd Conference and Workshop on Intrusion Detection and Response*, San Diego, CA, SANS Institute, 1999.

[DAH 12] DAHL G.E., YU D., DENG L. *et al.*, "Context-dependent pre-trained deep neural networks for large-vocabulary speech recognition", *IEEE Transactions on Audio, Speech, and Language Processing*, vol. 20, no. 1, pp. 30–42, January 2012.

[DIT 01] DITTRICH P., ZIEGLER J., BANZHAF W., "Artificial chemistries – a review", *Artificial Life*, vol. 7, no. 3, pp. 225–275, 2001.

[ENA 10] ENAMI N., MOGHADAM R.A., DADASHTABAR K. *et al.*, "Neural network based energy efficiency in wireless sensor networks: a survey", *International Journal of Computer Science & Engineering Survey (IJCSES)*, vol. 1, no. 1, pp. 39–55, August 2010.

[FRE 10] FREEDMAN D.H., *Brainmakers: How Scientists Moving Beyond Computers Create Rival to Human Brain*, Touchstone, 2010.

[GEL 01] GELENBE E., LENT R., XU Z., "Measurement and performance of a cognitive packet network", *Computer Networks*, vol. 37, pp. 691–791, 2001.

[GHO 98] GHOSH A.K., WANKEN J., CHARRON F., "Detecting anomalous and unknown intrusions against programs", *Proceedings of the Annual Computer Security Application Conference*, pp. 259–267, 1998.

[HAW 05] HAWKINS J., BLAKESLEE S., *On Intelligence*, St. Martin's Griffin, 2005.

[HAY 99] HAYKIN S., *Neural Networks: A Comprehensive Foundation*, 2nd ed., Macmillan College Publishing Company, NY, 1999.

[HEA 08] HEATON J., *Introduction to Neural Networks for Java*, 2nd ed., Heaton Research, Inc., 2008.

[HOP 82] HOPFIELD J.J., "Neural networks and physical systems with emergent collective computational abilities", *Proceedings of the National Academy of Sciences*, vol. 79, pp. 2554–2558, April 1982.

[HUE 08] HUEBSCHER M.C., MCCANN J.A., "A survey of autonomic computing – degrees, models, and applications", *ACM Computing Surveys*, vol. 40, no. 3, August 2008.

[HUY 01] HUYCK C.R., "Cell assemblies as an intermediate level model of cognition, emergent neural computational architectures based on neuroscience", *Lecture Notes in Computer Science*, vol. 2036, pp. 383–397, 24 July 2001.

[JAG 14] JÄGER G., ZUG S., BRADE T. *et al.*, "Assessing neural networks for sensor fault detection", *2014 IEEE International Conference on Computational Intelligence and Virtual Environments for Measurement Systems and Applications (CIVEMSA)*, pp. 70–75, 5–7 May 2014.

[JAN 89] JANIG W., "Autonomic nervous system", in SCHMIDT A., THEWS G. (eds.), *Human Physiology*, 2nd ed., Springer-Verlag, New York, NY, pp. 333–370, 1989.

[KAN 03] KANDEL E.R., "The molecular biology of memory storage: a dialog between genes and synapses", in JÖRNVALL H. (ed.), *Nobel Lectures, Physiology or Medicine 1996-2000*, World Scientific Publishing Co., Singapore, 2003.

[KEP 03] KEPHART J.O., CHESS D.M., "The vision of autonomic computing", *IEEE Computer*, vol. 36, no. 1, pp. 41–50, 2003.

[KOC 99] KOCH C., LAURENT G., "Complexity and the nervous system", *Science*, vol. 284, no. 5411, pp. 96–98, April 1999.

[KOH 82] KOHONEN T., "Self-organized formation of topologically correct feature maps", *Biological Cybernetics*, vol. 43, no. 1, pp. 59–69, 1982.

[KUL 09] KULKARNI R.V., VENAYAGAMOORTHY G., "Neural network based secure media access control protocol for wireless sensor networks", *Proceedings of the International Joint Conference on Neural Networks*, June 2009.

[LIP 00] LIPPMANN R.P., CUNNINGHAM R.K., "Improving intrusion detection performance using keyword selection and neural networks", *Elsevier Computer Networks*, vol. 34, no. 4, pp. 597–603, October 2000.

[MCC 43] MCCULLOCH W.S., PITTS W., "A logical calculus of ideas immanent in nervous activity", *Bulletin of Mathematical Biophysics*, vol. 5, no. 4, pp. 115–133, 1943.

[MIN 69] MINSKY M., PAPERT S., *An Introduction to Computational Geometry*, MIT Press, 1969.

[MOL 14] MOLNÁR Z., KAAS J.H., CARLOS J.A. *et al.*, "Evolution and development of the mammalian cerebral cortex", *Brain Behaviour and Evolution*, vol. 83, no. 2, pp. 126–139, 24 April 2014.

[NAS 07] NASSER N., GUIZANI S., AL-MASRI E., "Middleware vertical handoff manager: a neural network-based solution", *Proceedings of the 2007 IEEE International Conference on Communications (ICC '07)*, Glasgow, Scotland, pp. 5671–5676, June 2007.

[NIE 98] NIEUWENHUYS R., DONKELAAR H.J., NICHOLSON C., *The Central Nervous System of Vertebrates*, Springer, 1998.

[NUG 02] NUGENT A., Physical neural network design incorporating nanotechnology, Patent US 20030177450 A1, 12 March 2002.

[PAK 97] PAKKENBERG B., GUNDERSEN H.J., "Neocortical neuron number in humans: effect of sex and age", *Journal of Comparative Neurology*, vol. 384, no. 2, pp. 312–320, 28 July 1997.

[PAS 12] PASTORINO E., PORTILLO S.D., *What Is Psychology? Essentials*, 2nd ed., Cengage Learning, 2012.

[PEN 50] PENFIELD W., RASMUSSEN T., *The Cerebral Cortex of Man*, Macmillan, New York, 1950.

[PET 84] PETERS A., JONES E.G. (eds), *Cerebral Cortex*, Springer Science+Business Media, LLC, 1984.

[PRZ 08] PRZYSTALKA P., "Model-based fault detection and isolation using locally recurrent neural networks", *Lecture Notes in Computer Science*, vol. 5097, pp. 123–134, 2008.

[RAU 88] RAUCH H.E., WINARSKE T., "Neural networks for routing communication traffic", *IEEE Control Systems Magazine*, vol. 8, no. 2, pp. 26–31, April 1988.

[ROJ 96] ROJAS R., *Neural Networks: A Systematic Introduction*, Springer, 1996.

[ROS 58] ROSENBLATT F., "The perceptron: a probabilistic model for information storage and organization in the brain", *Psychological Review*, vol. 65, no. 6, pp. 386–408, 1958.

[RYA 97] RYAN J., LIN M.J., MIIKKULAINEN R., Intrusion detection with neural networks, Technical Report WS-97-07, AAAI, 1997.

[SCH 93] SCHOTT G.D., "Penfield's homunculus: a note on cerebral cartography", *Journal of Neurology, Neurosurgery, and Psychiatry*, vol. 56, no. 4, pp. 329–333, April 1993.

[SCO 57] SCOVILLE W.B., MILNER B., "Loss of recent memory after bilateral hippocampal lesions", *Journal of Neurology, Neurosurgery and Psychiatry*, vol. 20, no. 1, pp. 11–21, 1957.

[SHA 08] SHAHBAZI H., ARAGHIZADEH M.A., DALVI M., "Minimum power intelligent routing in wireless sensors networks using self organizing neural networks", *International Symposium on Telecommunications 2008 (IST '08)*, August 2008.

[SIR 08] SIRCAR S., *Principles of Medical Physiology*, Thieme, 2008.

[SIV 03] SIVANANDAM P., *Introduction to Artificial Neural Networks*, Vikas Publication House Pvt Ltd, 2003.

[STE 03] STERRIT R., BUSTARD D., "Autonomic computing – a means of achieving dependability?", *Proceedings of the 10th IEEE International Conference and Workshop on the Engineering of Computer-Based Systems (ECBS '03)*, IEEE, 2003.

[SWE 06] SWENSON R.S., Review of clinical and functional neuroscience, Dartmouth Medical School, available at http://www.dartmouth.edu/~rswenson/NeuroSci/index.html, 2006.

[TES 96] TESSIER-LAVIGNE M., GOODMAN C.S., "The molecular biology of axon guidance", *Science Magazine*, vol. 274, no. 5290, pp. 1123–1133, 15 November 1996.

[THO 96] THORPE S.J., FIZE D., MARLOT C., "Speed of processing in the human visual cortex", *Nature*, vol. 381, pp. 520–522, 1996.

[TVE 01] TVETER D.R., *The Backprop Algorithm*, available at: www.dontveter.com/bpr/bpr.html, 2001.

[WER 74] WERBOS P.J., Beyond regression: new tools for prediction and analysis in the behavioral sciences, PhD Thesis, Department of Applied Mathematics, Harvard University, 1974.

[WID 61] WIDROW B., PIERCE W.H., ANGELL J.B., Birth, life, and death in microelectronic systems, Technical Report No. 1552-2/1851-1, Solid-State Electronics Laboratory, Stanford University, Stanford, CA, 1961.

[WIK 14] WIKIBOOKS CONTRIBUTORS, Human physiology, available at http://en.wikibooks.org/wiki/Human_Physiology, 2014.

[WU 10] WU S.X., BANZHAF W., "The use of computational intelligence in intrusion detection systems: a review", *Applied Soft Computing*, vol. 10, no. 1, pp. 1–35, January 2010.

[ZHA 14] ZHANG W., LIU K., ZHANG W. *et al.*, "Wi-Fi positioning based on deep learning", *2014 IEEE International Conference on Information and Automation (ICIA)*, July 2014.

4

Swarm Intelligence (SI)

*Word cloud representing the full text of this chapter and
the words frequencies. Created with Wordle.net*

Nature is amazing and it provides inspiration for computer scientists in the most distinct ways. *Swarm intelligence* is a good example of this. As Gestalt psychologist Kurt Koffka once said, "The whole is other than the sum of the parts". Swarm intelligence goes along the same line of thinking; it is linked to group intelligence behavior, how smart the collective behavior is and how the behavior of each individual member contributes to the whole. Sometimes, the synergy between members of one group, e.g. the collective behaviors of social insects such as ants and bees, presents an intelligence that goes far beyond the intelligence of each of its members. Some researchers even consider the collective, the swarm, as one single and unique individual. This individual can perform actions, or exhibit a behavior, that the individual members of the group alone would not be able to do. The main characteristic of swarm intelligence algorithms is that they automatically evolve by stimulating the social behavior of organisms. By sharing information and taking the information of others into account, the behavior of the individuals is optimized to achieve a certain objective, which is normally much higher than the individual actions of the members of the group.

However, swarm intelligence is not the only system where the behavior of the whole is different from, or cannot be simply explained by, the behavior of the parts. This kind of pattern is quite common in nature; for example, each biological organism seems to exemplify this concept, the clearest one being our own brain. Our intelligent behavior is much more complex and much richer than the simple work performed by our individual neurons. One would hardly classify the human brain as swarm intelligence. Corne *et al.* [COR 12] argue that to be classified as swarm intelligence, the biological inspiration has to (1) present a behavior that emerges from the cooperation of a group of (2) largely homogeneous individual agents, (3) each one performing simple tasks. Such agents act in parallel (4) in an asynchronous way (5) and with (6) no or little central coordination. The communication among agents also has to be the (7) result of stigmergy.

Stigmergy, from the Greek στιγμα (stigma), meaning "mark" or "sign", and εργον (ergon), meaning "work/action", is a mechanism of spontaneous and indirect coordination between agents, where traces left in the environment, resulting from one action, encourage the performance of a subsequent action by the same or a distinct agent. Stigmergy promotes self-organization, as it fosters efficient collaboration between simple agents, who lack any memory, intelligence or even individual awareness of each other [MAR 08]. The French biologist Pierre-Paul Grassé introduced the term in an essay about the work of termites [GRA 59]. Stigmergy is now considered one of the key concepts in the field of swarm intelligence [PAR 03]. Swarm intelligence concerns the cooperation of individual agents to achieve a defined and precise goal, even if the agents are unaware of such collaboration. Agents interact with the environment and with other agents as a way to overcome their individual cognitive limitation.

Different biological systems can be classified as being swarm intelligent, e.g. bird flocking, fish schooling and bacterial colonies. Insect societies, for example, are capable of impressive tasks, such as bees finding the best source of nectar within the range of their hive, ants finding the shortest path between their nests and sources of food. We tend to anthropomorphize insects, and other social animals, attributing to them human characteristics such as diligence and altruism, as they put the good of the colony above their own needs. However, biological swarm systems are far from the School of Fish, giving information to Marlin and Dory in *Finding Nemo*, the Pixar film. The actions of the individual agents are much more reflexive acts than conscientious ones. In fact, each agent executes quite simple, most often repetitive tasks, reacting to the stimuli of the immediate environment. The global cooperation emerges as the result of a genetic pre-programmed behavior, where individuals perform the same actions according to the same set of stimuli. Moreover, performing the procedures, the individuals change the environment, which creates a new set of stimuli for them and all the others in the same region. This, in turn, triggers another set of actions. However, these are hardly meticulously planned actions.

Termites, for example, use pheromones to build their complex nests by following a simple decentralized rule set and by changing the environment they are in. Each insect makes a mud ball, impregnates with pheromones and places it, first, at random over a target region. The probability of depositing a mud ball at a given location increases with the sensed presence of other mud balls and the sensed concentration of pheromones. Termites are attracted by the pheromones of individuals of the same colony; so the natural tendency is to accumulate the mud at the same places and, the larger the pile of mud, the more attractive the spot. The few initial random placements increase the other termites' probability of deposing their balls on those places. Over time, this leads to the construction of columns, and eventually these columns meet at the top to form arches and after that tunnels and chambers.

In [REY 87], Reynolds describes a series of rules to emulate the group behavior of a group of animals. The "Reynold rules" were developed in the computer graphics community, and they present a simple yet efficient way to emulate the natural swarm-like behavior of groups of animals. Since their proposition, these rules have become a standard for the film industry and have been used in a series of different movies. The rule parameters are adapted to the species that is being modeled, but the triplet of steering behaviors to be followed by each individual in a swarm consists of:

– *separation*: steer to avoid coming too close to others;

– *alignment*: steer toward the mean heading of others;

– *cohesion*: steer toward the mean position of others.

Figure 4.1. *Nearly 2 m tall termite mound, Minas Gerais, Brazil*

However, where can we use this kind of technique? What are the problems we can solve with them? In [PAR 03], Parunak argues that swarm techniques can be applied to problems that have the D4 characteristics: diverse, distributed, decentralized and dynamic.

By "diverse", it means that these techniques can treat a broad range of information types, coming from diverse information sources. Swarm-based methods have been used from image synthesis to routing on wireless networks. The data sources go from temperature sensors to high-resolution video analyses.

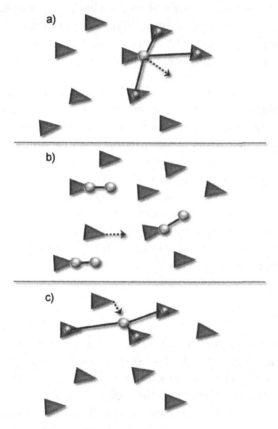

Figure 4.2. *Illustrating Reynolds' rules [REY 87], the work that served as inspiration for the particle swarm optimization technique. a) The separation where the velocity of the birds is adjusted according to the flock mates, b) the alignment where each bird adjusts its direction toward the average, the average of those in its perceptual field and c) cohesion shows the direction adjustment, where the bird moves it toward the mean position of the flock mates in its perceptual field*

Two of the most influential works in the field, and which in some sense started it, are the works of Dorigo *et al.* on ant system optimization [DOR 91] and the work of Kennedy and Ebhart [KEN 95] on particle swarm optimization (PSO), based on birds flocking. Not only are they among the first works on the swarm field, but they are also quite representative of the two main families of swarm intelligence tactics. The first is characterized by a pheromone trail-following behavior, inspired, for example, by the behavior of ants, i.e. ant colony optimization (ACO) algorithms. The second family of methods is inspired by birds flocking and insects swarming, i.e. particle swarm optimization.

In general, we use bio-inspired techniques when we have to solve difficult optimization problems, in general with a multi-objective function, i.e. the problem that has more than one objective. Moreover, normally, these objectives are conflicting with each other and hardly any solutions for the problem exist. In these cases, we are interested in finding a good trade-off between the different solutions. The set of desired solutions is the one that represents the best possible compromises among the various objectives, which is called a Pareto front optimization [YAN 01]. Swarm optimization techniques have shown to be extremely efficient in finding solutions for multi-objective optimization problems.

4.1. Ant colony optimization

ACO was first introduced by Dorigo *et al.* in 1991. The algorithm, at that time, was called *ant system* (AS), and is still quite representative of the works in the field. There are literally thousands of different species of ants and their diets vary widely; they may be carnivorous, herbivorous or omnivorous. The ones that serve as an inspiration to the ACO are the ones that search for food and bring it back to the colony. Just for the sake of precision, some kinds of ants do not actually eat the leaves they collect. Instead, they use them to feed a fungus, and it is this fungus that they eat. In any case, we are interested in the behavior of the ants that forage in groups.

When an ant from the group finds food, it marks the trail back to the colony with a pheromone. This trail is then followed by other ants, which, in their turn, also reinforce the pheromone trail on their way back. When that specific source is exhausted, ants stop marking it and, with time, the trail is forgotten as the pheromone dissipates. If new sources of food are found, and the route to it is shorter than the previous route, the trail is marked and ants that follow this path will take less time to go and return. For this reason, the scent on this new path will slowly increase and the old, longer path with time will be forgotten. The reinforcement will act over the shorter routes; what ants are doing, in a distributed way, is finding the shortest path to the food sources [GOS 89].

The ACO algorithm closely follows the behavior of natural ant colonies. Artificial ants lay down artificial pheromone trails along the route they take to solve the problem at hand. The next agents create their own solutions, but are influenced by the behavior of their predecessors, using these artificial pheromone trails.

4.2. Applications on networks

4.2.1. *Ants colony on routing*

The first initiative to apply ACO algorithms on wireless networks is the GPS/ant-like routing algorithm (GPSAL), the work of Câmara and Loureiro [CAM 00]. The approach is simple and quite effective in disseminating routing information among the nodes. Each node on the network has the right to generate ant agents at random intervals and at random nodes in the network. These ants follow different paths spreading routing information about the previous nodes to the following nodes on the route. When arriving at the destination, the ant agent is sent back, possibly by another path, spreading the routing information about all the nodes it has passed before. When arriving back to the source node, the node that issued the ant has access to the most recent information possible about the nodes in the ant path.

The routing information has a timestamp that is used to distinguish between older and newer information. The routing method is a "flexible" source routing, i.e. the full path of the packet is determined on the origin and inserted into the header of the packet. However, intermediate nodes may deviate the ant if they have better/fresher routes to the destination. The intermediate node collects the information about node position on the header of the ant, and may update the information on the header of the ant packet if they have more recent information about a given node. Stigmergy arises from the interaction between the ant packets and the nodes. Each ant benefits from the knowledge acquired from the previous ants that passed on the node, and helps to improve the knowledge of the node to help the next ants that will pass by. Moreover, from time to time, nodes exchange tables with their neighbors, which helps even more to spread the information over the network.

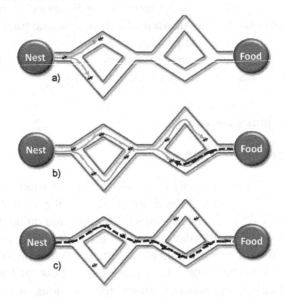

Figure 4.3. *Foraging ants, in the beginning a) search at random for a source of food, with time the smaller paths start to get a stronger pheromone scent b) and finally c) almost all ants start to use the shortest path. Inspired by the work of Goss et al.*

Considering a mobile *ad hoc* network, where nodes use a multi-hop communication strategy, i.e. they trust their neighbors to rebroadcast their messages to the next node in the path, the ant agents carry the most updated information possible about the path they just followed. It is important to note that, in general, mobile *ad hoc* networks are considered without a centralized structure. There is no central node to organize the network. GPSAL is fully distributed and the random ant message distribution is responsible for disseminating routing updated information and providing an up-to-date view of the network to each node. This, of course, comes with a cost, the overhead of handling the ant messages. However, GPSAL never uses flooding to spread information. Flooding is a common but expensive technique to spread information on wireless *ad hoc* networks. It consists of spreading the information to all nodes in the network, or in a controlled flooding to a subset of nodes. Each node that receives a message forwards it to all other nodes that it is connected to, apart from that which it received the message from. It is simple as it does not make assumptions about the type of nodes, e.g. position or traffic pattern, and flooding is quite effective in spreading information. At one moment or another, the large majority of the routing algorithms for *ad hoc* networks may resort to the use of flooding. Unfortunately, flooding is extremely expensive in terms of messages generated, thus in energy and medium use. Even in the configuration where GPSAL generates the maximum number of ants it is much less expensive, in terms of generated messages, than a flooding based algorithm.

In the ant-based control (ABC) [SCH 97] scheme, the routing tasks are divided into two: exploratory ants which make probabilistic decisions and the actual data routing mechanism that is the deterministic application of the routing over the data packet. The chosen path is always the one that presents the highest level of pheromone. Exploratory ants are used for route updates. Each source node sends a number of exploratory ants that are forwarded to a randomly selected destination. The routing table at each node stores a normalized amount of pheromone for each other node on the network; the neighbors are stored as rows and all possible destinations are stored as columns. The amount is normalized; for each column, the sum amount of pheromone is 1; thus, it can be used as probabilities for selecting the best link. Exploratory ants make their choices based on a random number generation to select one of these probabilities. Ants walk through the nodes of the network one node per time step; the age of the ant when it reaches the destination reflects the length of its path. If a number of such ants are released in a source node and follow different routes through a given destination, then the youngest ant at the arrival corresponds to the best path.

While being forwarded, the ant packets are unaware of how long it will take to reach the destination. However, it knows very well how long it took to reach here from the source node. The pheromone trails on ABC make reference not only to the destination but also to the source node. The amount of pheromone released is a function of the age of the ant, i.e. its distance from the source, and it is used to update the table for the source in the nth row that represents the neighbor that forwarded the ant.

The data packets always choose the path with the highest level of pheromone, but do not leave any. Both ants and calls travel on the same queue. Calls make a deterministic choice of a link with the highest probability, but do not leave any pheromone.

In the original ABC work, exploratory ant agents update only the table entry corresponding to the source of the ant. However, as an optimization, Guerin [GUE 97] proposes that while moving forward, the table for all nodes where the ant passed should be updated as well. The cost for this would be to append the identificator of the node on the ant packet before each forwarding.

4.2.2. Ants colony on intrusion detection

With the increase in the availability of online systems, the intrusion detection system (IDS) has quickly become one of the most important elements of the network security infrastructure. An intrusion is an attempt to use system resources without privileges, possibly causing incidental damage. Intrusion detection is the process of monitoring the events occurring in a computer system or network, analyzing them and warning about possible incidents.

When attacking a machine, intruders often exploit multiple vulnerabilities to acquire control of the system. A network attack graph (NAG) [NOE 04] is a way to describe the possible network attack scenarios. By analyzing the NAG, we can find the minimum set of exploits that can provide access to the machine. AntNag [ABA 06] is an ant colony-based optimization algorithm for the minimization analysis of NAGs. The AntNag algorithm searches for different attacks on a NAG, where each edge represents an exploit and every complete path from an initial node to a target node corresponds to an attack scenario. The minimization analysis of the graph gives the minimum set of exploits that should be eliminated to ensure that none of the represented attacks is feasible. Unfortunately, this problem is proved to be NP-hard [SHE 02]; thus, to solve it in a timely way for large attack graphs, heuristic approaches are required.

The vulnerabilities graph is created using the nodes' connectivity, network goals and the vulnerabilities detected by vulnerability scanning tools. Based on the NAG, a number of ants iteratively construct a set of critical exploits by incrementally adding exploits until all attack scenarios are covered. Systematically, each ant incrementally constructs a critical set of exploits. The exploits are modeled as edges on the graph, and the choice of which path the ant will follow is based on the amount of pheromone associated with the edges of the graph. The higher the level of pheromone, the bigger the probability of an ant following that path. While constructing a possible solution, ants also update the pheromone trails. However, differently from what happens with real ants, on AntNag, ants decrease the amount of pheromone on the path when taking it. The consequence is that when one ant takes a path, it lowers the probability of the next ants following that edge of the graph.

Considering the path of the ants, it is also possible that the chosen set of exploits is not minimal, i.e. it presents redundant exploits. Another extra task to eliminate redundant exploits from the ant's critical set is also considered.

In [FEN 01], Fenet and Hassas propose an architecture for a distributed stealth intrusion detection and response system (IDRS). The idea is to use the ant colony metaphor to locate the source of an attack. The detection of an intrusion triggers the release of an alert pheromone through the network and the agents, sensitive to this chemical message, converge to its source point to initiate a defensive action. IDRS is based on a number of sedentary agents called pheromone servers installed on each host and that provide the interface between the mobile agents and the pheromone information collected and stored in each machine. Pheromone servers are responsible for spreading an alert-like message throughout the network, performing the evaporation of the pheromone and controlling the access to the pheromone information to only authorized agents. The watcher is a static agent installed on each host that monitors the processes of that host and its network connections. It is the real detection mechanism, where the whole intelligence of detecting an attack resides. In the case of an attack, the watcher releases an alert that summarizes and identifies the attack. After that, other agents, lymphocytes, are created to constrain the attack. This part of the system takes its basic inspiration from immune systems, but the fast and efficient distribution of information over the network is inspired by the ant colony heuristic.

The intrusion detection based on emotional ants for sensors (IDEAS) algorithm [BAN 05] proposes an IDS that uses emotional ants to track intruders' trails and detect the source of attacks. This system relies on agents that are located on different nodes to monitor the activities of the host, its neighbors and the traffic on the search for evidence of possible attacks. The network is modeled as a graph where ant-like agents are randomly distributed over the nodes and traverse the connection graph. Nodes choose their path based on the pheromone trail, which represents the possible attacks. Only the ant that generated the best tour since the beginning of the trail, i.e. closer to detecting an attack, is allowed to globally update the pheromone concentration on the branches. As the example of IDRS on this approach when ants traverse a given edge, the local pheromone update rule decreases the level of the pheromone on that edge. This intends for ants to avoid converging all along the same path.

The emotional ants are efficient adaptive agents that, guided by the pheromone, can evaluate variable conflict tendencies to adjust their forwarding schema. However, the emotional ants do not base their movement only on the pheromone trails. They also consider "social relations" with other ants, based on the affective computing theory [PIC 97]. The proposed emotional model is primarily based on certain thematic reactions, such as affinity, satisfaction dejection and approach, exhibited by the emotional ant agents.

The self-organized ANT colony-based intrusion detection system (ANTIDS) [RAM 05] proposes a method where the ant agents form clusters by pick and drop items, with certain probability. ANTIDS is, in fact, a data mining method that tries to organize features that represent attacks into clusters. On this approach, Ramos and Abraham argue that instead of having agents exploring the network randomly, which can be inefficient since they may go to uninteresting regions, it is better to have the agents following the pheromone concentrations. The two major factors that influence the actions of the ant are the number of objects in its neighborhood and their similarity. The picking and dropping of an object is based on Lumer and Faieta's method [LUM 94]. ANTIDS uses the local information of a number of objects, and their similarity, to control the picking and dropping of objects. According to Ramos and Abraham, this has a series of advantages. First, it allows ants to find clusters of objects in an adaptive manner. Second, it eliminates the need for memory in the agents, which makes the scheme less demanding in terms of resources. Finally, it speeds up the algorithm on the search for optimal solutions, as the ants tend to move to areas of higher interest.

4.3. Particle swarm optimization

The PSO method was proposed by Kennedy and Eberhart [KEN 95] in 1995. It uses as inspiration the swarm-like behavior described in the works of Reynolds [REY 87] and Heppner and Grenander [HEP 90]. The basic idea behind PSO is to relate the behavior of a flock of birds moving in three-dimensional (3D) space to a swarm of solutions over an optimization problem. The birds' flocking flight follows three basic rules: (1) collision avoidance, birds tend to readjust their position to avoid neighbor birds' positions (2); velocity matching, particles tend to synchronize their speed with that of their neighbors; and (3) flock centering, which dictates individuals stay close to flock mates.

The same way the flock of birds moves, the particles move through a multi-dimensional search space toward good solutions. Each particle has a position and a velocity. The position represents the present solution, while the velocity is a vector that describes the navigation on the search space and that, hopefully, will lead to an improvement in the result for the next iteration. The information for the particle includes knowledge gained from its previous experience and knowledge gained from the swarm. In the beginning, the PSO is initialized with a group of random particles and then these particles search for the optimal value by following the best solutions found up to that moment. While searching for good solutions, each particle is considered a solution for the given problem. The value of the particle, which is estimated by the objective function, is used to update its information and to optimize the objective of the swarm. The most important part of the PSO is the calculation of the direction to take. The way this decision is made is heavily inspired by the work of Reynolds [REY 87]. However, the implementation details are different and PSOs have the capacity to perform the fitness evaluation of the present solution considering local and global values. The collective experiences are evaluated and, after evaluation, the particles choose the next movement to take in the search space.

The influences that lead to the position update are [COR 12]:

– *current velocity*: the particle's current velocity;

– *personal best*: the particle remembers the best fitting arrangement it has encountered. This value is considered on the calculation of the direction for the next move;

– *global best*: particles in the swarm are aware of the overall best position that any particle has found. This position is considered on the calculation of the direction to go from the present position.

The basic update rule for the velocity of the next iteration (*t+1*) is given by:

$$v_i(t+1) = \omega v_i(t) + c_1 r_1 (p_i - x_i) + c_2 r_2 (g - x_i) \qquad [4.1]$$

where ω is a constant that represents the inertia weight, c_1 and c_2 are the acceleration constants, r_1 and r_2 are the random numbers, p_i is the personal best position of particle i, g is the global best position among all particles in the swarm and x_i is the current position of particle i. The update rule for the next position ($x_i(t+1)$) is given by:

$$x_i(t+1) = x_i(t) + v_i(t+1) \qquad [4.2]$$

where $x_i(t)$ is the current position and $v_i(t+1)$ is the result of equation [4.1].

4.4. Applications on networks

4.4.1. *Particle swarm on node positioning*

The node deployment problem refers to determining the best position for network nodes so that the connectivity and coverage are maximized while, normally, other constraints such as energy efficiency are considered as well. This kind of problem is well studied in the framework of wireless sensor networks, where the node concentration may be high and the energetic efficiency is a major issue to improve the network lifetime. A number of PSO-based approaches have been proposed to solve the positioning problem in wireless sensor networks. An interesting comparison of some of these techniques can be found in [KUL 11].

Aziz *et al.* [AZI 07] proposed an offline approach based on PSO and the Voronoi partition algorithm [VOR 08]. The main objective of the algorithm presented in [AZI 07] is to minimize the uncovered areas on the sensed region. The area is divided into a grid and the sensors should cover these regions. The PSO–Voronoi circumvents these regions involving the sensors with Voronoi polygons. The PSO particles are the sensors' positions; for each particle, a set of Voronoi polygons is determined. The cost function is the number of areas of the grid that are uncovered by the sensors.

Li *et al.* in [LI 07b] study the use of particle swarm and genetic optimization (PSGO) over the nodes position for improving nodes coverage problem in a network composed of a mix of mobile and static nodes. The PSGO method rearranges the mobile nodes positions to improve the average nodes density. This proposition uses the concepts of mutation and selection from genetic algorithms to improve the global PSO performance. At each iteration, PSGO discards the worst particles and generates new particles at random locations. Moreover, some particles move at random locations. Even if not fully in line with the standard PSO, these mechanisms clearly improve the coverage of the search space. In [LI 09], Li and Lai also, inspired by genetic algorithms, proposed to change randomly the direction of some particles to avoid the problem of premature convergence.

The problem of underwater sensor deployment and sensing area coverage is studied in light of PSO in [DU 14]. Underwater sensor deployment is an interesting problem. Not only we need to place sensors in a 3D space, but also we need to take into account the particular dynamics of the water. The particle swarm inspired underwater sensor deployment (PSSD) algorithm controls the motion of the sensors, considering not only the direction calculated through the traditional PSO method, but also taking into account the water flow to try to save the node energy over the sensor movement.

4.4.2. *Particle swarm on intrusion detection*

Intrusion detection is a difficult problem and, in general, PSO-based IDSs are normally hybrid systems, where PSO is used in combination with other machine-learning techniques. The intrusion detection method proposed by Michailidis *et al.* [MIC 08] uses PSO to train an artificial neural network (ANN). During the training phase, PSO is executed recursively to calibrate the weights of the ANN. Then, the ANN makes a classification of the possible attacks. The particles in this case represent the multi-dimensional vectors composed of the ANN parameters. The particle with the optimum adaptation values is the one chosen to feed the ANN. In the same line of thinking, Tian *et al.* [TIA 10] also proposed an IDS that combines PSO and ANN. The PSO is used as a preprocessor for the ANN. The role of the neural network is to select the subset of attributes to be evaluated, and the PSO is used to optimize the neural network parameters and improve its performance.

Another technique that is often used with PSO is support vector machines (SVMs) [BOS 92]. SVM is a statistical learning method that presents a good learning and generalization capacities, even in high-dimensional and noisy environments. An IDS that uses SVM and PSO is the improved binary particle swarm optimization with support vector machine (IBPSO-SVM) [LIU 10]. The method also uses rough set theory (RST) [LI 07a] to decrease the number of parameters by subtracting redundant and noisy attributes. After applying RST, the PSO is applied to optimize parameters of the SVM and increase its classification accuracy. The velocity of the particles is updated using a linear decreasing weight [SHI 98] technique to accelerate even more the speed of the PSO convergence. In IBPSO-SVM, the main role of the PSO is to optimize the parameter selection. In [SRI 07], the same combination of PSP and SVM is used. In a first phase, the training data are preprocessed to handle missing and incomplete data inputs. Then, the remaining features are evaluated and selected using PSO. The SVM performs the clustering for detecting possible intrusions.

Wang *et al.* [WAN 09] propose the use of PSO to search the optimal SVM parameters and extract a representative feature subset. Each particle represents a solution that indicates which feature and parameter values should be kept. The best particle, along with the training data set, is used as an input or the SVM classifier, which tries to classify specific network behavior as intrusive or normal.

Xiao *et al.* [XIA 06] proposed an IDS that is based on PSO and K-means [MAC 67]. In this proposal, each particle's position represents the set of D-dimensional centroids produced by the K-means algorithm. Thus, each particle's position is represented as an array. For each particle, the PSO fitness function evaluates the present position and updates the particle array. Then, the K-means algorithm optimizes the new generation of particles. The algorithm shows a high convergence speed.

In [ALJ 13], the authors present a parallel intrusion detection system (IDS-MRCPSO) based on the MapReduce framework. MapReduce [DEA 04] is a parallel processing model especially adapted to data-intensive applications. The proposal treats the intrusion detection as an optimization problem, and uses PSO to solve this problem. Analyzing large network traffic data to detect intrusions may be data intensive and takes a long time; therefore, the PSO is parallelized using the MapReduce model [DEA 04] to scale even when considering high network traffic. In the classification part, the authors use individual centroids to represent different clusters. First, these centroids are chosen randomly from the data set; however, iteratively, they are updated based on the particles velocities, and convergence is made to a global best value.

The MapReduce task consists of mapping each particle as a value and the particle identification number as a key. The map function retrieves the map values, which contain the entire particle's information. The centroid vector is updated using the previous information and the PSO equations. At the end, the map function issues the updated centroid to the reduce function. The reduce function uses the key to organize the intermediate map output that is used by the next fitness MapReduce module. The fitness evaluation takes into account the distances between all data records and the particle centroid vector. The map function in this module treats each record as a value and each record's identification number as a key. For each particle, the map extracts the information in the centroid and returns it to the reduce function. The reduce function groups the values with the same key and the minimum distances for each particle centroid. The reduce function groups the values with the same key and the minimum distances for each particle centroid. This is afterwards used as a new fitness value for the centroid. The final global best centroid vector will be chosen as the detection model.

4.5. Further reading

CORNE D.W., REYNOLDS A., BONABEAU E., "Swarm intelligence", *Handbook of Natural Computing*, Springer, pp. 1599–1622, 2012.

PARPINELLI R.S., LOPES H.S., "New inspirations in swarm intelligence: a survey", *International Journal of Bio-Inspired Computation*, vol. 3, no. 1, pp. 1–16, 2011.

KOLIAS C., KAMBOURAKIS G., MARAGOUDAKIS M., "Swarm intelligence in intrusion detection: a survey", *Computer Security*, vol. 30, no. 8, pp. 625–642, November 2011.

SALEEM M., CARO G.A.D., FAROOQ M., "Swarm intelligence based routing protocol for wireless sensor", *Elsevier Information Sciences*, vol. 181, pp. 4597–4624, 2011.

REN H., MENG M.Q.H., "Biologically inspired approaches for wireless sensor networks", *Proceedings of IEEE International Conference on Mechatronics and Automation*, 2006.

4.6. Bibliography

[ABA 06] ABADI M., JALILI S., "An ant colony optimization algorithm for network vulnerability analysis", *Iranian Journal of Electrical & Electronic Engineering (IJEEE)*, vol. 2, no. 3, pp. 106–120, 2006.

[ALJ 13] ALJARAH I., LUDWIG S.A., "MapReduce intrusion detection system based on a particle swarm optimization clustering algorithm", *Proceedings of the 2013 IEEE Congress on Evolutionary Computation*, Cancún, México, 20–23 June 2013.

[AZI 07] AZIZ N.A.B.A., MOHEMMED A.W., SAGAR B.S.D., "Particle swarm optimization and Voronoi diagram for wireless sensor networks coverage optimization", *Proceedings of the International Conference on Intelligent and Advanced Systems (ICIAS '07)*, pp. 961–965, 2007.

[BAN 05] BANERJEE S., GROSAN C., ABRAHAM A. *et al.*, "Intrusion detection in sensor networks using emotional ants", *International Journal of Applied Science and Computations*, vol. 12, no. 3, pp. 152–173, 2005.

[BOS 92] BOSER B.E., GUYON I.M., VAPNIK V.N., "A training algorithm for optimal margin classifiers", *Proceedings of the 5th Annual Workshop on Computational Learning Theory*, New York, NY, ACM Press, pp. 144–152, 1992.

[CÂM 00] CÂMARA D., LOUREIRO A.A.F., "A GPS/ant-like routing algorithm for *ad hoc* networks", *IEEE Wireless Communications and Networking Conference (WCNC '00)*, Chicago, IL, September 2000.

[COR 12] CORNE D.W., REYNOLDS A., BONABEAU E., "Swarm intelligence", *Handbook of Natural Computing*, Springer, pp. 1599–1622, 2012.

[DEA 04] DEAN J., GHEMAWAT S., "MapReduce: simplified data processing on large clusters", *OSDI'04: Sixth Symposium on Operating System Design and Implementation*, San Francisco, CA, December, 2004.

[DOR 91] DORIGO M., MANIEZZO V., COLORNI A., The ant system: an autocatalytic optimizing process, Technical Report No. 91-016 Revised, Politecnico di Milano, Italy, 1991.

[DU 14] DU H., XIA N., ZHENG R., "Particle swarm inspired underwater sensor self-deployment", *Sensors*, MDPI, vol. 14, pp. 15262–15281 2014.

[FEN 01] FENET S., HASSAS S., "A distributed intrusion detection and response system based on mobile autonomous agents using social insects communication paradigm", *Proceedings of the 1st International Workshop on Security of Mobile Multi agent Systems (SEMAS)*, pp. 41–48, 2001.

[GOS 89] GOSS S., ARON S., DENEUBOURG J.L. *et al.*, "Self-organized shortcuts in the argentine ant", *Naturwissenschaften*, vol. 76, pp. 579–581, 1989.

[GRA 59] GRASSÉ P.P., "La reconstruction du nid et les coordinations interindividuelles chez Bellicositermes natalensis et Cubitermes sp. la théorie de la stigmergie: Essai d'interprétation du comportement des termites constructeurs", *Insectes Sociaux*, vol. 6, no. 1, pp. 41–80, 1959.

[GUE 97] GUERIN S., Optimisation multi-agents en environment dynamique: Application au routage dans les réseaux de télécommunications, DEA Dissertation, University of Rennes I, France, 1997.

[HEP 90] HEPPNER F., GRENANDER U., "A stochastic nonlinear model for coordinated bird flocks", in KRASNER S. (ed.), *The Ubiquity of Chaos*, AAAS Publications, Washington, 1990.

[KEN 95] KENNEDY J., EBERHART R., "Particle swarm optimization", *Proceedings of IEEE International Joint Conference on Neural Networks*, IEEE Press, Piscataway, pp. 1942–1948, 1995.

[KUL 11] KULKARNI R.V., VENAYAGAMOORTHY G.K., "Particle swarm optimization in wireless-sensor networks: a brief survey", *IEEE Transactions on Systems, Man, and Cybernetics, Part C: Applications and Reviews*, vol. 41, no. 2, pp. 262–267, March 2011.

[LI 07a] LI D.Y., HUI B.Q., "A kind of dynamic rough sets", *Proceedings of the 4th International Conference on Fuzzy Systems and Knowledge Discovery*, vol. 3, pp. 79–85, August 2007.

[LI 07b] LI J., LI K., WEI Z., "Improving sensing coverage of wireless sensor networks by employing mobile robots", *IEEE International Conference on Robotics and Biomimetics (ROBIO '07)*, pp. 899–903, 15–18 December 2007.

[LI 09] LI Z., LEI L., "Sensor node deployment in wireless sensor networks based on improved particle swarm optimization", *Proceedings of 2009 IEEE International Conference on Applied Superconductivity and Electromagnetic Devices*, Chengdu, China, 25–27 September 2009.

[LIU 10] LIU H., JIAN Y., LIU S., "A new intelligent intrusion detection method based on attribute reduction and parameters optimization of SVM", *Proceedings of the 2nd International Workshop on Education Technology and Computer Science (ETCS)*, pp. 202–205, 2010.

[LUM 94] LUMER E.D., FAIETA B., "Diversity and adaptation in populations of clustering ants", *Proceedings of SAB'94 – 3rd Conference on Simulation of Adaptive Behavior: From Animal to Animats*, The MIT Press/Bradford Books, Cambridge, MA, 1994.

[MAC 67] MACQUEEN J.B., "Some methods for classification and analysis of multivariate observations", *Proceedings of 5th Berkeley Symposium on Mathematical Statistics and Probability*, pp. 281–297, 1967.

[MAR 08] MARSH L., ONOF C., "Stigmergic epistemology, stigmergic cognition", *Cognitive Systems Research*, vol. 9, nos. 1–2, pp. 136–149, March 2008.

[MIC 08] MICHAILIDIS E., KATSIKAS S.K., GEORGOPOULOS E., "Intrusion detection using evolutionary neural networks", *Proceedings of the Panhellenic Conference on Informatics 2008 (PCI '08)*, 2008.

[NOE 04] NOEL S., ROBERTSON E., JAJODIA S., "Correlating intrusion events and building attack scenarios through attack graph distances", *20th Annual Computer Security Applications Conference*, pp. 350–359, 6–10 December 2004.

[PAR 03] PARUNAK H.V.D., "Making swarming happen", *Proceedings of the Conference on Swarming and Network Enabled Command, Control, Communications, Computers, Intelligence, Surveillance and Reconnaissance (C4ISR)*, McLean, VA, January 2003.

[PIC 97] PICARD R.W., *Affective Computing*, MIT Press, Cambridge, MA, 1997.

[RAM 05] RAMOS V., ABRAHAM A., "ANTIDS: self organized ant based clustering model for intrusion detection system", *Proceedings of the 4th IEEE International Workshop on Soft Computing as Transdisciplinary Science and Technology (WSTST '05)*, pp. 977–986, 2005.

[REY 87] REYNOLDS C., "Flocks, herds and schools: a distributed behavioral model", *Computer Graphics*, vol. 21, no. 4, pp. 25–34, 1987.

[SCH 97] SCHOONDERWOERD R., HOLLAND O.E., BRUTEN J.L., "Ant-like agents for load balancing in telecommunication networks", *1st ACM International Conference on Autonomous Agents*, Marina del Ray, CA, 1997.

[SHE 02] SHEYNER O., HAINES J., JHA S. *et al.*, "Automated generation and analysis of attack graphs", *2002 IEEE Symposium on Security and Privacy*, pp. 273–284, 2002.

[SHI 98] SHI Y., EBERHART R.C., "A modified particle swarm optimizer", *Proceedings of the IEEE Congress on Evolutionary Computation*, Piscataway, NJ, pp. 69–73, 1998.

[SRI 07] SRINOY S., "Intrusion detection model based on particle swarm optimization and support vector machine", *Proceedings of the IEEE Symposium on Computational Intelligence in Security and Defense Applications (CISDA '07)*, April 2007.

[TIA 10] TIAN W., LIU J., "Network intrusion detection analysis with neural network and particle swarm optimization algorithm", *Proceedings of 2010 Chinese IEEE Control and Decision Conference (CCDC)*, pp. 1749–1752, 2010.

[VOR 08] VORONOI G., "Nouvelles applications des paramètres continus à la théorie des formes quadratiques", *Journal für die Reine und Angewandte Mathematik*, vol. 133, no. 133, pp. 97–178, 1908.

[WAN 09] WANG J., HONG X., REN R. *et al.*, "A real-time intrusion detection system based on PSO-SVM", *Proceedings of the International Workshop on Information Security and Application (IWISA '09)*, pp. 319–321, 2009.

[XIA 06] XIAO L., SHAO Z., LIU G., "K-means algorithm based on particle swarm optimization algorithm for anomaly intrusion detection", *Proceedings of the 6th World Congress on Intelligent Control and Automation (WCICA '06)*, pp. 5854–5858, 2006.

[YAN 01] YANG X., *Economics: New Classical Versus Neoclassical Frameworks*, Wiley-Blackwell, 2001.

Glossary

Allele: series of genes occupying a specific location on the chromosome and responsible for an individual trait, e.g. eye color.

Chromosome: thread-like structure within a cell composed of nucleic acids and proteins which stores the genetic material of the individual and carries the genetic information in the form of genes.

Dendrites: part of the neuron that is *connected* to other neurons and carries the impulse toward the cell body.

Ecosystem: a system that includes all living organisms in a given area, as well as the physical environment they are inserted into.

Entropy: the measurement of the amount of disorder in a system.

Epiphenomena: secondary effects or by-products, occurring simultaneously to a main phenomenon, but not directly related to it.

Fitness: the measure of the capacity of an organism to survive and reproduce in particular environment. It determines the impact of the individual in the genetic pool of the next generations.

Gamete: a reproductive cell that contains the haploid set of chromosomes, i.e. just half of the genetic material. Examples include spermatozoon (male) or ovum (female), reproductive cells.

Gene: the basic unity of hereditary information located in a chromosome. The information that determines the characteristics of the parent transferred to the offspring. It is a sequence of nucleotides along a segment of DNA with the coded instruction for the synthesis of RNA or a protein.

Genome: the complete set of genes of an organism.

Genotype: is the genetic expression of a specific trait, or set of traits, of an individual, or group of individuals. It is composed of a series of alleles and transmitted from the parent to the offspring.

Homeostasis: the ability a system has to maintain an equilibrium state, even in the presence of external perturbations.

Locus: the location of a gene, or sequence of genes, on a chromosome.

Mutation: random change in a gene that may lead to changes in the individual phenotype.

Nerve: a group of connected neurons working together to perform a similar task.

Neuron: a single nerve cell, it is the basic unit of the nervous system and is responsible for receiving, processing and transmitting messages.

Neurotransmitter: a chemical substance released by one neuron to excite a neighboring one. It is used to transmit nerve impulses from one neuron to another.

Nucleotide: the basic component of nucleic acids, such as DNA and RNA. A nitrogenous base, a sugar and a phosphate group make it up. Based on their structure, they can normally be divided into purines and pyrimidines. In DNA, the purine bases include adenine and guanine, while the pyrimidine bases are thymine and cytosine.

Pareto optimal: named after the Italian economist Vilfredo Pareto (1848–1923), it is an equilibrium allocation in which it is impossible to make any one of the parties better off without making at least one of the other participants worse off. It is considered a fair equilibrium point.

Phenotype: physical, or chemical, measurable characteristics of an organism, they express a trait of an organism and are the result of the individual genotype and its interaction with the environment.

Pheromone: a chemical substance that an animal produces, which changes the behavior of another animal of the same species.

Population: all individuals in an experiment.

Stigmergy: a mechanism of spontaneous, indirect coordination between agents, or actions, where the trace left in the environment by an action stimulates the performance of a subsequent action.

Stimulus: something that triggers a reaction, for example if someone puts his/her hand over a hot surface that is a stimulus, because the person will sense the temperature. Moreover, if the temperature is too high, it will trigger also the instinctive reaction of removing the hand away from it.

Synapse: a small gap between two nerve cells, where the impulses are transmitted through the help of a neurotransmitter.

Trait: possible aspect of an individual.

Index

Printed in the United States
By Bookmasters